METAL IONS IN BIOLOGICAL SYSTEMS

VOLUME 9

Amino Acids and Derivatives as Ambivalent Ligands

METAL IONS IN BIOLOGICAL SYSTEMS

Edited by

Helmut Sigel

Institute of Inorganic Chemistry
University of Basel
Basel, Switzerland

VOLUME 9
Amino Acids and Derivatives as Ambivalent Ligands

MARCEL DEKKER, INC.   New York and Basel

ISSN: 0161-5149

MARCEL DEKKER, INC.
270 Madison Avenue, New York, New York 10016

ISBN: 0-8247-6875-2
Library of Congress Catalog Card Number: 78-17297

Current printing (last digit):
10 9 8 7 6 5 4 3 2 1

PRINTED IN THE UNITED STATES OF AMERICA

PREFACE TO THE SERIES

Recently, the importance of metal ions to the vital functions of living organisms, hence their health and well-being, has become increasingly apparent. As a result, the long-neglected field of "bioinorganic chemistry" is now developing at a rapid pace. The research centers on the synthesis, stability, formation, structure, and reactivity of biological metal ion-containing compounds of low and high molecular weight. The metabolism and transport of metal ions and their complexes is being studied, and new models for complicated natural structures and processes are being devised and tested. The focal point of our attention is the connection between the chemistry of metal ions and their role for life.

No doubt, we are only at the brink of this process. Thus, it is with the intention of linking coordination chemistry and biochemistry in their widest sense that the series METAL IONS IN BIOLOGICAL SYSTEMS reflects the growing field of "bioinorganic chemistry." We hope, also, that this series will help to break down the barriers between the historically separate spheres of chemistry, biochemistry, biology, medicine, and physics, with the expectation that a good deal of the future outstanding discoveries will be made in the interdisciplinary areas of science.

Should this series prove a stimulus for new activities in this fascinating "field" it would well serve its purpose and would be a satisfactory result for the efforts spent by the authors.

Helmut Sigel

PREFACE TO VOLUME 9

Amino acids with a suitable side chain donor atom offer metal
ions this atom for coordination, as well as the amino and carboxylate
group. Depending on the structure of the polydentate amino acid or
derivative and the properties of the metal ion, all or only a part of
the potentially available donor atoms may coordinate, i.e., several
structural possibilities exist for the resulting complex. These
"ambivalent" or "ambidentate" qualities of amino acids and derivatives are in the center of this volume, while certain other properties of complexes with amino acids and related ligands have been
treated already in a number of chapters in Volumes 1, 2, and 5 of
this series.

The introductory chapter of the present volume covers complexes
with a number of amino acids with chelatable side chain donor atoms.
The coordinating properties of aspartic acid and glutamic acid,
cysteine and penicillamine, and L-dopa are outlined in separate
chapters. Other contributions examine the complexing properties
of reduced and oxidized glutathione, and the stereoselectivity in
the metal ion complexes of amino acids and dipeptides. The volume
terminates with a chapter on complexes of polypeptides, such as
corticotropin fragments. Metal ion-protein interactions have been
discussed previously in chapters of Volumes 3 and 6. Finally, it
may be added that the ligating ambivalency of nucleotides and derivatives has been covered in the preceding Volume 8.

Helmut Sigel

CONTENTS

CONTRIBUTORS

Numbers in parentheses indicate the pages on which the authors'
contributions begin.

K. BURGER, Institute of Inorganic and Analytical Chemistry,
L. Eötvös University, Budapest, Hungary (213)

C. A. EVANS, Department of Chemistry, The University of Alberta,
Edmonton, Alberta, Canada (41 and 103)

A. GERGELY, Department of Inorganic and Analytical Chemistry,
Kossuth Lajos University, Debrecen, Hungary (77 and 143)

R. GUEVREMONT, Department of Chemistry, The University of Alberta,
Edmonton, Alberta, Canada (41 and 103)

R. J. W. HEFFORD, Department of Inorganic and Structural Chemistry,
The University, Leeds, England (173)

T. KISS, Department of Inorganic and Analytical Chemistry,
Kossuth Lajos University, Debrecen, Hungary (143)

R. B. MARTIN, Department of Chemistry, University of Virginia,
Charlottesville, Virginia (1)

L. D. PETTIT, Department of Inorganic and Structural Chemistry,
The University, Leeds, England (173)

D. L. RABENSTEIN, Department of Chemistry, The University of
Alberta, Edmonton, Alberta, Canada (41 and 103)

I. SÓVÁGÓ, Department of Inorganic and Analytical Chemistry,
Kossuth Lajos University, Debrecen, Hungary (77)

CONTENTS OF OTHER VOLUMES

Other volumes are in preparation.

Comments and suggestions with regard to contents, topics, and the
like for future volumes of the series would be greatly welcome.

METAL IONS IN BIOLOGICAL SYSTEMS

VOLUME 9

Amino Acids and Derivatives as Ambivalent Ligands

Chapter 1

COMPLEXES OF α-AMINO ACIDS WITH CHELATABLE
SIDE CHAIN DONOR ATOMS

R. Bruce Martin
Department of Chemistry
University of Virginia
Charlottesville, Virginia

1. SCOPE

Of the 20 α-amino acids that commonly occur in proteins, about
half contain side chain donor atoms that are at least potentially
capable of forming a chelate ring with a metal ion bound at the
α-amino nitrogen. If the metal ion is also chelated by the α-amino
and α-carboxylate groups that occur in the free amino acid, then two
chelate rings will be formed. A listing of a variety of α-amino
acids with potentially chelatable side chains appears in Table 1.
The amino acids are grouped by the ring size formed by a chelate
between the α-amino nitrogen and the indicated donor atom on a side
chain. Other factors being equal, five-membered chelate rings are
favored, so that we anticipate a general weakening of the tendency
toward side chain chelation as ring size increases.

Several entries in Table 1 are the subject of other chapters in
this volume and thus are considered only briefly here. Complexes of
aspartic acid and glutamic acid, with β- and γ-carboxylate side
chains, are reviewed in Chap. 2. The strong metal ion binding

TABLE 1

Terdentate α-Amino Acids: Ring Size
and Donor Atom

Ring size	Donor atom		
	0	N	S
5	Serine (Ser) Threonine (Thr)	2,3-Diaminopropanoate (dap)	Cysteine (Cys) Cystine Penicillamine
6	Aspartate (Asp) Asparagine (Asn)	2,4-Diaminobutanoate (dab) Asparagine (Asn) Histidine (His)	Homocysteine Methionine (Met)
7	Glutamate (Glu) Glutamine (Gln)	Ornithine (Orn) Arginine (Arg)	
8		Lysine (Lys)	

engendered by the chelatable sulfhydryl groups of cysteine and penicillamine (β,β-dimethylcysteine) is discussed in Chap. 3. Homocysteine with a second -CH$_2$ group between the α carbon and the sulfhydryl group has been little studied for its metal ion binding capacities, but it is anticipated that it would be a weaker metal ion binder than cysteine. The coordination chemistry of a cysteinyl residue in the tripeptide glutathione is reviewed in Chap. 4.

Recently discovered in proteins such as prothrombin that are involved in the complex blood-clotting process, γ-carboxyglutamic acid possesses two carboxylate groups at the γ carbon in addition to the α-carboxylate group. The additional γ-carboxylate group arises by a postribosomal vitamin K-dependent carboxylation of an existing glutamate side chain in a protein. Although the γ-carboxyglutamic acid residues are essential for tight Ca^{2+} binding, they are insufficient without the requisite protein tertiary structure [1]. Because the coordination chemistry of a γ-carboxyglutamic acid residue in a protein is that of a substituted malonate, it is not considered further here.

Several commonly occurring α-amino acids do not appear in Table 1. The nitrogen atom of the indole ring of tryptophan does not participate in metal ion binding so that this amino acid is bidentate in its coordination chemistry. The hydroxy groups on hydroxyproline and tyrosine are sterically prohibited from forming a chelate with the glycinate locus and are only weak metal ion binders in their own right. Stronger metal ion binding occurs at the catecholate function of L-3,4-dihydroxyphenylalanine (L-dopa), but again this binding site is not chelatable to the glycinate locus. A pH dependent competition occurs between the glycinate and catecholate functions of L-dopa for Cu^{2+} [2] and other metal ions as reviewed in Chap. 5.

This chapter concerns side chain binding sites that are potentially chelatable to a metal ion already chelated at the glycinate locus in a five-membered ring. The emphasis herein is on complexes that occur in neutral solutions; complexes in acidic solutions with metal ions bound weakly only to carboxylate groups are not considered.

Most of the complexes occur as substituted glycinates and the question becomes one of the degree of interaction of the side chain with a metal ion already chelated at the glycinate locus. Side chain chelate ring sizes such as those in Table 1 are determined from the α-amino nitrogen. Because of the role of basicity in determining metal ion binding capability it is deemed essential to evaluate proton affinities of competing ligand binding sites. The chapter ends with a brief discussion of the role of aromatic side chains in complex formation (Sec. 9).

One of the important reasons for delineating amino acid ligand bonding modes to metal ions is that most of the non-protein-bound Cu^{2+} in the blood plasma is bound to amino acid ligands. Mixed Cu^{2+} complexes of histidine and cystine are estimated to account for almost half of low molecular weight plasma Cu^{2+}, the 2:1 histidine complex 11%, and mixed histidine complexes with about 14 other amino acid ligands for most of the remainder [3, 4]. The important subject of mixed ligand complexes is the concern of the entire Vol. 2 in this series.

In addition to stability constant comparisons, visible absorption and circular dichroism (CD) spectroscopy provide powerful insights into ligand bonding modes in complexes. This subject has been reviewed in Vol. 1 of this series [5]. Of special interest here is the ability of amino acid chelation modes, alternative to the substituted glycinate one, such as bidentate αNH_2, ωNH_2, and terdentate chelation, to be signaled by sign inversions in the visible CD of d-d transitions [6, 7]. Use of these evident changes in CD spectra will be made in interpreting bonding modes of the α,ω-diaminocarboxylates and histidine in Secs. 7 and 8. More subtle effects are also apparent, and the visible CD of mixed Cu^{2+} complexes has been employed to elucidate possible ligand-ligand interactions [8, 9].

The literature often reflects uncertainty as to whether side chain interactions occur in some of the weaker donor situations. Some of the differences are reduced if cognizance is taken of the physical methods employed. Interactions that do not contribute or contribute only marginally to the stability of a complex may still

be evident in some dynamic method of analysis such as line broadening
by paramagnetic metal ions in nuclear magnetic resonance (nmr) spectroscopy, which is sensitive to very weak interactions with only a small
fraction of the ligand present in a particular complex structure. We
shall attempt to limit the use of the term "weak interaction" to instances where the stability of a complex is increased a significant
amount over that anticipated if the interaction did not occur.

2. STABILITY CONSTANT COMPARISONS

In assessing relative metal ion binding strengths among a group of
ligands, it is difficult to allow for differences in ligand basicity.
Most often these differences are ignored. However, greater proton
affinity is expected to produce a greater metal ion binding capability.
The reaction $M^{2+} + HL \rightleftharpoons ML^+ + H^+$ portrays the competition between
proton and a metal ion for a ligand donor site, and its equilibrium
constant given by $\log K_1 - pK_a$ furnishes a realistic measure of
protonated ligand metal ion binding capability. Use of this equation
and its "conditional" equilibrium constant to allow for differences
in ligand basicity implies that a plot of $\log K_1$ vs. pK_a is linear
with unit slope and passes through the origin. This can be shown to
be false for most series of complexes. For the ligands of interest
in this article, it seems evident that use of the "conditional"
equilibrium constant overestimates the role of basicity in determining relative metal ion binding strengths. For example, the equilibrium constant for the above reaction with Cu^{2+} is further to the
right with Asn^- than with Asp^{2-}, yet most investigators would agree
that Asp^{2-} should be listed as at least as strong a metal ion binder
as Asn^-. In order to provide some acknowledgment of the role of
ligand basicity in determining relative metal ion binding strengths,
this chapter adopts for these ligands and metal ions a convenient
compromise guide of $\log K_1 - 0.7 pK_a$. Values of this difference
appear in parentheses after $\log K_1$ values in Table 2 and represent
basicity-adjusted stability constants. The less negative or more

TABLE 2

α-Amino Acid Acidity and Stability Constant Logarithms[a]

Ligand[b]	pK_a[c]	Co^{2+}			Ni^{2+}		
		$\log K_1$	$\log K_2$	$\log K_1/K_2$	$\log K_1$	$\log K_2$	$\log K_1/K_2$
Gly⁻	9.57	4.64 (−2.1)[d]	3.82	0.8	5.78 (−0.9)	4.80	1.0
Ala⁻	9.69	4.31 (−2.5)	3.50	0.8	5.40 (−1.4)	4.47	0.9
α-ab⁻	9.62	4.21 (−2.5)	3.50	0.7	5.35 (−1.4)	4.41	0.9
Ser⁻	9.08	4.38 (−2.0)	3.62	0.8	5.43 (−0.9)	4.53	0.9
Thr⁻	8.97	4.38 (−1.9)	3.63	0.8	5.45 (−0.8)	4.51	0.9
Gln⁻	9.00	4.04 (−2.3)	3.28	0.8	5.16 (−1.1)	4.26	0.9
Asn⁻	8.74	4.51 (−1.6)	3.50	1.0	5.68 (−0.4)	4.55	1.1
Glu²⁻	9.59	4.56 (−2.2)	3.30	1.3	5.60 (−1.1)	4.16	1.4
Asp²⁻	9.63	5.95 (−0.8)	4.28	1.7	7.16 (0.4)	5.24	1.9
Phe⁻	9.11	4.05 (−2.3)	3.51	0.5	5.15 (−1.2)	4.44	0.7
Met⁻	9.05	4.12 (−2.2)	3.44	0.7	5.19 (−1.2)	4.65	0.5
SMC⁻	8.73	4.12 (−2.0)	3.49	0.6	5.26 (−0.9)	4.56	0.7
dap°	6.66	2.91 (−2.0)			4.04 (−0.8)	3.48	0.6
dap⁻	9.39	6.28	5.08	1.2	8.16	7.00	1.2
dab°	8.15	3.40 (−2.3)			4.54 (−1.2)	3.79	0.8
dab⁻	10.20	6.75	5.25	1.5	8.91	7.06	1.9
Orn°	8.75	3.65 (−2.5)	3.09	0.6	4.73 (−1.4)	4.00	0.7
Orn⁻	10.52	5.01	3.48	1.5	7.11	4.92	2.2
Lys°	9.14	3.84 (−2.6)	3.23	0.6	4.93 (−1.5)	4.22	0.7
Lys⁻	10.67				5.75	4.59	1.2
Arg°	9.02	3.86 (−2.5)	3.05	0.8	4.99 (−1.3)	4.02	1.0
His⁻	9.08	6.90 (0.5)	5.44	1.5	8.67 (2.3)	6.87	1.8

[a]Most values taken from Ref. 27 at 25° and 0.1 M ionic strength and sometimes refined according to references given in text. Values for last six ligands except His taken from Refs. 78 and 86 which agree well with other values in Refs. 79 and 87.

[b]Nonstandard abbreviations used are: α-ab, α-aminobutyrate; SMC, S-methylcysteine; dap, 2,3-diaminopropanoate; and dab, 2,4-diamino-butanoate. For other abbreviations see Table 1 or the abbreviation list at the end of the chapter.

6

Cu^{2+}			Zn^{2+}			log K$_1$
log K$_1$	log K$_2$	log K$_1$/K$_2$	log K$_1$	log K$_2$	log K$_1$/K$_2$	Cu-Ni
8.15 (1.5)	6.88	1.3	4.96 (-1.7)	4.23	0.7	2.4
8.13 (1.4)	6.79	1.3	4.58 (-2.2)	4.00	0.6	2.7
8.07 (1.3)	6.75	1.3	4.50 (-2.2)	4.15	0.4	2.7
7.89 (1.5)	6.59	1.3	4.65 (-1.7)	4.03	0.6	2.5
8.01 (1.7)	6.72	1.3	4.67 (-1.6)	3.99	0.7	2.6
7.75 (1.5)	6.48	1.3				2.6
7.86 (1.7)	6.56	1.3				2.2
7.87 (1.2)	6.29	1.6	4.59 (-2.1)	3.66	0.9	2.3
8.57 (1.8)	6.78	1.8	5.84 (-0.9)	4.31	1.5	1.4
7.86 (1.5)	6.91	1.0	4.29 (-2.1)	4.06	0.2	2.7
7.87 (1.5)	6.85	1.0	4.37 (-2.0)	3.96	0.4	2.7
7.88 (1.8)	6.84	1.0	4.46 (-1.7)	4.06	0.4	2.6
6.16 (1.3)	5.12	1.0	3.20 (-1.7)	2.64	0.6	2.1
10.51	9.32	1.2	6.30	5.35	1.0	2.4
6.94 (1.2)	5.86	1.1	3.74 (-2.0)	3.34	0.4	2.4
10.50	8.52	2.0	6.70	5.60	1.1	1.6
7.29 (1.2)	6.12	1.2	3.73 (-2.4)	3.08	0.7	2.6
7.62 (1.2)	6.49	1.1	4.06 (-2.3)	3.46	0.6	2.7
7.56 (1.2)	6.45	1.1				2.6
10.2 (3.8)	7.9	2.3	6.55 (0.2)	5.51	1.0	1.5

[c]Concentration pK$_a$ values for ammonium group deprotonations. To convert to more usual mixed acidity constant logarithms add 0.11.

[d]Values in parentheses are log K$_1$ - 0.7 pK$_a$. See Sec. 2.

positive the numbers in parentheses the greater the basicity-
adjusted stability.

Three different tests of possible involvement of the side chain
in α-amino acid complexes are provided in Table 2. Terdentate chelation should appear as unusual values of one or more of the following
three parameters when compared to "normal" bidentate values for Ala$^-$
and α-aminobutyrate (α-ab$^-$), results for which are in close agreement.
As described in the previous paragraph, to allow for differences in
ligand basicity the difference $\log K_1 - 0.7 pK_a$ appears in parentheses after $\log K_1$ values. Under each metal ion, the last column
lists the difference between the logarithms of the first and second
stability constants, $\log K_1 - \log K_2$. Representative differences
for a wide variety of Cu^{2+} complexes range from 0.5 to 3.0, and some
interpretations of the results are presented [10]. Finally, the last
column of Table 2 lists the difference between the first stability
constants for Cu^{2+} and Ni^{2+} complexes, $\log K_1(Cu^{2+}) - \log K_1(Ni^{2+})$.
It is important to restrict comparisons to closely related ligands
such as those listed in Table 2. Even the first entry Gly$^-$ displays
some deviations from Ala$^-$ that, did we not know better, might lead
to its classification as a weakly terdentate ligand.

A cursory inspection of deviations from the values of Ala$^-$ for
the three terdentate test parameters suggests that Orn^0, Lys^0, and
Arg^0 are uniformly bidentate, Gln$^-$ and Glu^{2-} display a very weak
terdentate capability with Co^{2+} and Ni^{2+}, whereas for all metal ions
the second Glu$^-$ is more weakly bound than usual due to electrostatic
repulsion and is probably only bidentate, Asn$^-$ is terdentate with
Co^{2+} and Ni^{2+} and more weakly so with Cu^{2+}, Ser$^-$ and Thr$^-$ are similar
to Asn$^-$ only weaker, and Asp^{2-} is terdentate and even to some extent
with Cu^{2+}, where again the second ligand is mainly bidentate. Of the
thioether ligands, SMC$^-$ appears weakly terdentate and Met$^-$ less so
if at all. Outright terdentate ligands with $\log K_1 - 0.7 pK_a$ values
more than one log unit greater than the other ligands of Table 2 are,
in order of increasing binding strength, His$^-$, dab$^-$, and dap$^-$. Other
aspects of some of the many comparisons possible within Table 2 are
mentioned during the course of this chapter.

3. ASPARAGINE AND GLUTAMINE

As just mentioned in the previous section, the γ-amide carbonyl oxygen of Gln$^-$ appears to interact very weakly with Co^{2+} and Ni^{2+} but not Cu^{2+} bound at the glycinate locus. For Asn$^-$ the basicity-adjusted relative binding strengths in parentheses in Table 2 indicate considerable chelation of the β-amide carbonyl oxygen with Co^{2+} and Ni^{2+} and less so with Cu^{2+}. When allowance is made for basicity differences, Asn$^-$ forms weaker complexes than Asp^{2-} with Co^{2+} and Ni^{2+}, but complexes of comparable strength with Cu^{2+}.

Results of visible CD spectra support chelation of the amide carbonyl oxygen of Asn$^-$ to Ni^{2+} [11, 12] and suggest a weaker similar chelation to Cu^{2+} [6]. Crystal structures of bis complexes of Asn$^-$ with Cu^{2+}, Zn^{2+} [13], and Cd^{2+} [14] show only bidentate chelation and an interaction of the amide carbonyl oxygen with a metal ion in another complex.

Titration of a solution containing Cu^{2+} and excess Asn turns from blue to a more violet color in basic solutions [6]. Deprotonations occurring with pK_a 10.5 and 12.0 [15] correspond to substitution of Cu^{2+} for an amide hydrogen on each of two ligands and binding at the amide nitrogen. The absorption maximum at 580 nm indicates that four nitrogen donors, two amino and two deprotonated amides, on two Asn dianions chelate about the Cu^{2+} tetragonal plane [6]. Amide hydrogen substitution does not take place with Gln and Cu^{2+}. With the more strongly binding and tetragonal enPd(H$_2$O)$_2^{2+}$ substitution of an amide hydrogen occurs with Asn with pK_a 6.4 and with Gln with pK_a 9.0 [16]. In the resulting complexes the amino acids are coordinated through amino and deprotonated amide nitrogen donors. Thus a six- and even a seven-membered chelate ring involving nitrogen donors does form with Pd^{2+}. If Pd^{2+} alone was used as a criterion, Gln^{2-} would appear in the seven-membered ring N donor atom category in Table 1.

4. SERINE AND THREONINE

Whether the β-alcohol groups of Ser and Thr participate in chelation
has been a matter of disagreement in the literature. As indicated
by the basicity-adjusted binding strengths in Table 2, a weak chelation is indicated to the hexacoordinate metal ions Co^{2+}, Ni^{2+}, and
Zn^{2+}, whereas little chelation is evident by this test with Cu^{2+}.
These conclusions are in general agreement with a stability and
enthalpy comparison study, which, however, infers a stronger alcohol
chelation to Cu^{2+} [17]. Below pH 4.8 weak Cu^{2+} complexes of net
zero-charged Ser^0 and Thr^0 occur, presumably due to chelation of
alcohol and carboxylate groups [18]. Existence of these amino-
protonated complexes cannot be taken to imply binding of the alcohol
groups of anionic Ser^- and Thr^- to Cu^{2+}, as the stronger α-aminocar-
boxylate chelation now occupies the tetragonal plane and axial binding is weak. Visible CD spectra support the substituted glycinate
mode of coordination of Ser^- and Thr^- to Cu^{2+} [19, 20]. Broadening
of the methyl protons of excess Thr at about pH 8 is greater for
hexacoordinate Mn^{2+} than for tetragonal Cu^{2+}, which induces little
broadening [21]. Taken together these results in solution indicate
some chelation of Ser^- and Thr^- alcohol groups to hexacoordinate
metal ions and notably weaker chelation to Cu^{2+}. Crystal structure
determinations show only substituted glycinate modes of chelation
without alcohol coordination in complexes of Ser^- with Ni^{2+} [22],
Cu^{2+} [23], and Zn^{2+} [24], and of Thr^- with Cu^{2+} [25].

In basic solutions Cu^{2+} promotes the ionization of the alcohol
hydrogens of two chelated Thr^- with pK_a 10.3 and 11.3 [20]. After
ionization each of the Thr^{2-} are terdendate with glycinate and alco-
holate group chelation. The visible CD of the L-Thr-Cu^{2+} complexes
now develops positive peaks characteristic of terdentate chelation
that are absent in the predominantly negative CD spectra of L-amino
acids chelated only as substituted glycinates.

Cu^{2+} complexes of the dipeptides Ser-Gly and Thr-Gly (and
SMC-Gly) are considered to have enhanced stabilities prior to deprotonation of the amide hydrogen due to apical interaction of the side

chains [26]. Primary Cu^{2+} binding is at the amino nitrogen, and it is debatable as to which of the peptide carbonyl oxygen or the side chain alcohol oxygen (or ether sulfur) take up equatorial and axial positions. The stability constant for Cu^{2+} with ethanolamine [27] is greater than that with the two dipeptides (as is the stability constant for Cu^{2+} with 2-methylthioethylamine [27] greater than that with SMC-Gly). In any case terdentate chelation to Cu^{2+} of these amino terminal side chains prior to peptide deprotonation seems established.

After peptide hydrogen ionization, when the Cu^{2+} is tetragonally chelated by amino and deprotonated peptide nitrogens and a carboxylate oxygen, apical interaction of the β-alcohol oxygen (and β-ether sulfur) in the side chains is suggested for both Thr-Gly- and Gly-Thr-type dipeptide complexes [26]. This proposed apical interaction is not evident in visible CD where the sign and magnitude are typical of noninteracting side chains [19]. After peptide deprotonation the reinforced planarity of the amide bond and tetragonal terdentate chelation results in a stiffened peptide backbone in a complex and reduces the opportunity for a side chain-axial interaction in nonamino terminal residues [19].

5. METHIONINE AND S-METHYLCYSTEINE

The ether sulfur is an extremely weak base undergoing protonation only in strongly acid solutions with a pK_a of -6.8 [28], more than four log units weaker than values for oxygen ethers (-2.5) [29] and ethanol (-1.9) [30]. Even though it is closely situated to the chelating α-amino acid function of Met and SMC, the weak basicity of the thioether does not provide a basis for its coordination. Sulfur atoms are soft bases and interact most favorably with class b [31] or soft [32] metal ions. It is this softness that furnishes the basis for interaction of thioethers with metal ions. We anticipate that the ether sulfur will interact best with soft metal ions, marginally with borderline metal ions, and not at all with hard metal ions.

Comparisons among the basicity-adjusted constants in Table 2
suggest that there is a quite weak side chain interaction in SMC$^-$
that is comparable to that in Ser$^-$ for Co^{2+}, Ni^{2+}, and Zn^{2+} whereas
the Cu^{2+} interaction is stronger with SMC$^-$ but still quite weak.
Interaction of the ether sulfur in Met$^-$ is weaker still. Most borderline metal ions consistently form slightly weaker complexes with
both Met$^-$ and SMC$^-$ than with Ala$^-$. Lack of an appreciable chemical
shift of the methyl protons in nmr spectroscopy in solutions of SMC
and Zn^{2+}, Cd^{2+}, and Pb^{2+} (in contrast with Hg^{2+}) indicates little
involvement of the thioether group [33]. These results suggest that
the thioether group does not contribute importantly to complex stability in Met and SMC, both of which chelate borderline metal ions
primarily as substituted glycinates.

On the other hand, modest stability constant enhancements are
observed when some of the above metal ions complex to other thioether-
containing ligands of carboxylates [34, 35] and of 2-methylmercapto-
ethylamine [36] when compared to ammonia of similar pK_a. A general
review of metal ion-thioether interactions of biological interest
appears in Vol. 1 of this series [34]. Comparison of the coordination
tendencies of the thioether group as part of bidentate and terdentate
ligands suggests to us a greater tendency for the thioether group to
be evident as one donor atom of a bidentate ligand of lesser stability
constant than as the third donor atom in a terdentate ligand of greater
stability constant. This tendency is augmented for Cu^{2+}, where the
more favored donor atoms occupy tetragonal positions and the ether
sulfur is forced to interact at an apical position if at all.

Though the thioether group does not contribute importantly to
the stability of Met and SMC with hard or borderline metal ions,
there are suggestions that they may interact weakly or part of the
time with the ether sulfur. In solutions containing a 10^3:1 ratio
of Met or SMC to Cu^{2+} or Mn^{2+}, at a pH more than a log unit below
the pK_a of the ammonium group, appreciable broadening of the methyl
protons in the nmr spectra occurs only for SMC with Cu^{2+} [34, 37].
Even in this case, however, the broadening is not nearly as great as
for the hydrogen bound to the α carbon. The broadening in both cases

is scalar-determined, so that it is not possible to draw firm conclusions without knowledge of the hyperfine coupling constants at each ligand nucleus. Directly related to the distance from the paramagnetic ion, proton T_{1P} measurements show, under conditions of excess ligand, no significant thioether involvement with Mn^{2+} in Met and SMC [38] nor with Cu^{2+} in Met [39]. A weak interaction between the ether sulfur and Cu^{2+} occurs in SMC. Chelation of Cu^{2+} at the glycinate locus and a weak interaction of the thioether group in an apical position account for the observations [39]. That the interaction is weak (and/or apical) is supported by the absence of a reported absorption in the 400-nm region due to complexation of a thioether group in the tetragonal plane of Cu^{2+} [40, 41].

Solution studies on Met and SMC leave little doubt as to the modes of binding of a wide variety of metal ions. It should therefore be gratifying to crystallographers that the crystals harvested for structure determinations do indeed reflect predominant structures in solution even for labile metal ions. Thus, as expected, the thioether group is not coordinated to borderline metal ions in Met complexes of Cu^{2+} [42], Zn^{2+} [43], Cd^{2+} [44], Gly-Met complex of Cu^{2+} [45], and SMC complexes of Zn^{2+} and Cd^{2+} [46]. On the other hand, the ether sulfur and amino nitrogen are both coordinated in the Cu^{2+} tetragonal plane in a bis complex of 2-methylmercaptoethylamine [47]. The contrast provided by a bound thioether in the bidentate ligand offers additional support for the conclusion advanced in the last two sentences of the third paragraph of this section.

Interpretations of infrared spectra of solid complexes of Met [48, 49] and SMC [50] are in general but not always specific accord with solution and crystal structure investigations.

Whether the thioether group of Met and SMC is bound to inert Co(III) appears to depend on other factors, some of which are kinetically determined. The ether sulfur is coordinated in bis complexes of Met [51] and SMC [52]. With only two sites available in $en_2Co(SMC)^{2+}$, the thioether group is unbound in SMC, which chelates as a substituted glycinate [53].

The degree of interaction of metal ions with the thioether group of Met and SMC provides a test of their class b or softness character. On the basis of the results presented, both Co^{3+} classified as hard and Cd^{2+} classified as soft [32] are more aptly described as borderline. Co^{3+} binds the thioether group in Met and SMC, and for Cd^{2+} solution studies and crystal structure determinations of both Met and SMC reveal little or no significant interaction with the ether sulfur.

There is little doubt that the class b or soft-metal ions Pd^{2+}, Pt^{2+}, Ag^{2+}, Hg^{2+}, and CH_3Hg^+ coordinate to the ether sulfur of Met and SMC. Absorption spectra of solutions indicate sulfur binding in bis complexes of Met and SMC with Pd^{2+} [7]. A trans bis Met⁻ complex of Pt^{2+} with bidendate S and N chelation has been characterized [54]. Crystal structures also show bidentate S and N chelation to Pd^{2+} for both Met [55] and SMC [56] and to Pt^{2+} for Met [57]. Both nitrogens and the ether sulfur are bound to Pt^{2+} in a terdentate Gly-Met complex [57]. This result stands in contrast to a terdentate Gly-Met complex of borderline Cu^{2+} where a carboxylate oxygen replaces the ether sulfur [45].

Binding of a metal ion to an ether sulfur introduces asymmetry at the tetrahedral sulfur atom where three different substituents and a lone pair occur. If the ligand already contains an asymmetric carbon as in L-SMC or L-Met, metal ion bonding at the sulfur creates diastereomers. For many metal ions inversion at the sulfur atom will be rapid, but for Pt^{2+} the inversion process is slow enough to observe separate peaks in nmr spectra. Doubling of proton nmr peaks due to formation of diastereomers has been observed in a Pt^{2+} complex of SMC [58]. Inversion rates have been studied in many Pt^{2+} ether sulfur complexes [59].

Ag^+ coordinates without the benefit of chelation to the thioether of tetrahydrothiophene with a stability constant log K_S = 3.5 [60]. Proton nmr also reveals that the binding of Ag^+ and Hg^{2+} to Met and SMC is pH-dependent [33]. For both ligands Hg^{2+} shifts the methyl proton peak to a greater extent in acidic solutions, a result interpreted as exclusive Hg^{2+} binding to the thioether group when Hg^{2+} is

still unable to displace an ammonium proton from the more strongly
coordinating glycinate locus. Consistent with this tendency, a
crystal structure shows Hg^{2+} bound solely to the thioether of methionine with a protonated ammonium group [61]. As demonstrated by the
pH dependence of chemical shifts in ^1H nmr spectroscopy, CH_3Hg^+ binds
to Met exclusively at the thioether group at pH < 2 with a stability
constant of 87 [62]. By pH 7 to 8 the binding is solely at the amino
group with $\log K_s$ = 7.6. In accordance with these solution results,
a crystal prepared by mixing equimolar amounts of CH_3HgOH and dipolar
ion Met shows binding of CH_3Hg^+ at the amino nitrogen without thioether coordination [63].

 Crystallographic investigations reveal that the methionine
ether sulfur serves as a donor atom in naturally occurring metalloproteins. The heme iron of cytochrome c is also bound to a methionine sulfur [64]. In the intensely blue plastocyanin Cu^{2+} resides
in a distorted tetrahedral environment with side chain Met, Cys, and
two His donors [65]. Added class b or soft heavy-metal ion reagent
labels for protein crystallography are often found bound at Met sulfur.

6. CYSTINE

It has been suggested that the disulfide bond is an even weaker base
than an ether sulfur [29] (Sec. 5). Probably because cystine is
nearly insoluble in neutral aqueous solutions, only a few investigations with metal ions have been carried out. Mercurous acetate
simultaneously reduces and mercurates disulfide bonds to give RSHgSR
[66]. Estimates of stability constants have been used to suggest a
Cu^{2+} interaction with the disulfide bond of cystine [67]. There is
electron spin resonance evidence for an exchange interaction between
two Cu^{2+} in 2:1 frozen solutions at -196° with L-cystinyl-bis-glycine
that has been interpreted as due to binding of each Cu^{2+} to a different sulfur of the disulfide group [68]. No evidence for such binding
to the same ligand at room temperature was evident in an absorption
and CD study [69]. The last investigation reports spectral evidence

for a possible Cu^{2+}-disulfide interaction in 2:1 solutions with
L-cystinediamide. The coordination chemistry of the disulfide bond
in oxidized glutathione is reviewed in Chap. 4.

A crystal structure reveals no disulfide-Co^{3+} bonding in a
binuclear complex of L-cystine [70]. Possible weak disulfide bonding
to the sixth coordination position about Cu^{2+} occurs in a binuclear
D-penicillamine disulfide complex [71]. In crystal structures of
other aliphatic disulfide complexes, disulfide-metal ion bonding
occurs with Ni^{2+} [72] and Cu^+ [73-75]. Solid complexes of L-cystine
and several metal ions have been viewed by vibrational spectroscopy
[76].

7. α,ω-DIAMINOCARBOXYLATES

7.1. Protonation

It is instructive to consider features of both proton and metal ion
binding as a terminal amino group is displaced further out along a
growing alkyl chain from the glycinate portion of a molecule. The
series of four α,ω-diaminocarboxylates consists of 2,3-diaminopro-
panoate (dap), 2,4-diaminobutanoate (dab), 2,5-diaminopentanoate or
ornithine (Orn), and 2,6-diaminohexanoate or lysine (Lys). The most
basic carboxylate group of the four compounds in the series, that of
lysine, exhibits an acidity constant with pK_a 2.3, so that throughout
the series the carboxylate group is in its basic form in the pH region
of our interest. Table 3 lists the pK_a values for the second (pK_2)
and third (pK_3) deprotonations from each of the compounds of the
series. The usual mixed pK_2 and pK_3 values at 25° and 0.1 M ionic
strength quoted in Table 3 are derived from three papers [77-79],
the results of which are in excellent agreement when converted to a
common basis for ionic strength and in regard to mixed and concentration acidity constants. The deprotonation represented by pK_2 is commonly assigned to the α-ammonium group and that of pK_3 to the ω-
ammonium group. In the following analysis we show that this assumption is partially correct.

TABLE 3

Acidity Constants and Molar Ratio R of
[αNH_2, ωNH_3^+] to [αNH_3^+, ωNH_2] Forms

	$pK_2{}^a$	$pK_3{}^a$	$\log S^b$	$R_c{}^c$	$R_n{}^d$
2,3-Diaminopropanoate	6.77	9.51	2.21	1.0	1.1
2,4-Diaminobutanoate	8.27	10.31	1.18	5.0	
Ornithine	8.86	10.65	0.68	10.8	7.6
Lysine	9.25	10.79	0.33	14.2	9.8

[a]Usual mixed constants at 25° and 0.1 M ionic strength derived from Refs. 77–79.

[b]From Ref. 27; see text.

[c]Calculated as indicated in Sec. 7.1.

[d]Calculated from microconstants determined by nmr [84].

There are two pathways for deprotonation of the net positively charged α,ω-diaminocarboxylates to give the anionic form, the usually assumed route via the αNH_2, ωNH_3^+ species and the alternative pathway via the αNH_3^+, ωNH_2 species.

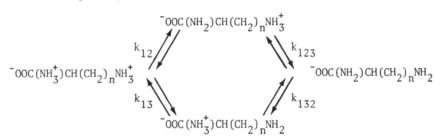

The value of n takes on values from 1 to 4 for dap through Lys. The subscripts 1, 2, and 3 are assigned to the carboxylic acid, α-NH_3^+, and ω-NH_3^+ groups, respectively. The last subscript on the microscopic acidity constants in the diagram refers to the group undergoing deprotonation. The macroscopic acidity constants determined by potentiometric titration (K_2 and K_3) are related to the microscopic acidity constants by the equations

$$K_2 = k_{12} + k_{13} \qquad K_3^{-1} = k_{123}^{-1} + k_{132}^{-1}$$

$$K_2 K_3 = k_{12} k_{123} = k_{13} k_{132}$$

There are only three independent equations relating the four unknown microconstants. An additional item of information is required to determine the molar concentration ratio $R = [\alpha NH_2, \omega NH_3^+]/[\alpha NH_3^+, \omega NH_2] = k_{12}/k_{13}$ of the two forms of net zero charge [80-82].

The reciprocal effects of the two ammonium group deprotonations on each other may be described by the ratio $S = k_{12}/k_{132} = k_{13}/k_{123}$ [82]. The more separated the two ammonium groups, the more nearly the ratio S approaches unity. Values of $\log S$ for the successive compounds in Table 3 may be estimated from the difference of the two acidity constant pK_a values [27] for 1,2-diaminoethane, 1,3-diaminopropane, 1,4-diaminobutane, and 1,5-diaminopentane, respectively, by means of the equation $\log S = pK_2 - pK_1 - 0.60$, where the 0.60 is a statistical factor. Values of $\log S$ so derived under the same conditions of 25° and 0.1 M ionic strength as the pK_2 and pK_3 values are listed in the fourth column of Table 3.

The relation among the macroscopic acidity constants and the two ratios R and S is given by $(R + 1)^2/R = K_2/K_3 S$ [83]. Appropriate values of R_c calculated from this equation appear in the fifth column of Table 3. For dap the calculated value of $pK_3 - pK_2 - \log S = 0.53$ is less than the lowest theoretical value of 0.60. Evidently for dap with closely spaced charges the $\log S$ value overestimates the degree of ammonium group interaction, probably because of significant conformational changes upon introduction of a carboxylate group into 1,2-diaminoethane. The result, however, does indicate a value of R_c near unity and this value is listed for dap in the fifth column of Table 3. The calculated values R_c may be compared with those determined from a chemical shift analysis in nmr spectroscopy [84] and listed as R_n in the sixth column of Table 3. It was not possible to resolve the chemical shifts for dab to provide a value for R_n. The two sets of R values listed in the fifth and sixth columns of Table 3 were determined by two entirely different methods. The agreement

between the two sets of R values is good especially because of the
sensitivity of the high R ratios found for Orn and Lys to even a few
hundredths of a log unit in pK_a values. The differences between the
R_c and R_n values for Orn and Lys constitute only 0.12 to 0.13 log
units in the several acidity constants used in the analysis. The
range of R values from 7.6 to 14.2 corresponds to 88 to 93% of the
αNH_2, ωNH_3^+ form. Availability of numerical values for R permits
calculation of each of the four microconstants by utilizing the
equations given above.

We conclude from the values of $R = [\alpha NH_2, \omega NH_3^+]/[\alpha NH_3^+, \omega NH_2]$
in Table 3, that for these two species of net zero charge the per-
centages of the αNH_2, ωNH_3^+ form are about only 50% for dap, 83% for
dab, and 90% for Orn and Lys. It is noteworthy that though the difference $pK_3 - pK_2$ is greatest for dap and least for Lys that the
ratio R is least for dap and greatest for Lys. As the equation for
R indicates, its value is determined not only by K_1/K_2 but also by
the interactions between the ammonium groups as represented by S.
For the compounds in Table 3, the ammonium group interactions, as
indicated by log S, increase 1.7 times more rapidly as n becomes
smaller than does the difference $pK_3 - pK_2$. The near equality of
the two ammonium group deprotonation pathways in dap is not shared
by its esters. The singly positively charged species of esters for
all four compounds in Table 3 exist predominantly in the αNH_2, ωNH_3^+
form.

In calculating the basicity-adjusted binding strengths for the
α,ω-diaminocarboxylates in Table 2, pK_1 acidity constants were employed in all cases except dap, where the concentration microconstant
value $pk_{12} = 6.96$ was used to allow for the equal basicities of the
α- and β-amino groups.

7.2. Complexation

The preceding analysis of protonation sites is additionally instructive because it points out a dilemma that also occurs in metal ion
binding to multidentate ligands. Though potentiometric titrations

can be analyzed to yield the amount and distribution of species of a
net charge type, such as the net-zero-charged species above, additional
arguments are required to specify the charge distribution within the
charge type, as the ratio R of $[\alpha NH_2, \omega NH_3^+]$ to $[\alpha NH_3^+, \omega NH_2]$ forms.
For metal ion complexes the additional arguments include stability
constant comparisons with related ligands (due allowance being made
for basicity differences among ligands) and any of several kinds of
spectroscopic results that provide information on the specific donor
groups complexed to a metal ion. In the cases of the α,ω-diamino-
carboxylates where there are alternatives between bidentate and terdentate chelation and between N and O donor atoms, visible absorption
and CD spectroscopy of transition metal ion complexes provide especially revealing additional information [6, 7], which any proposal
for probable binding modes is obligated to incorporate.

In solutions containing some excess ligand with Cu^{2+}, the predominant complex from pH 5 to 9 in solutions of Orn, Lys, and Arg is
$Cu(AH)_2^{2+}$ [79]. There is general agreement from stability constant
and spectroscopic arguments that these amino acids are coordinated
as substituted glycinates with protonated terminal ω functions [6,
78, 79]. Accordingly, the crystal structure of the bis-Cu^{2+} complex
of Orn^0 shows a substituted glycinate with a remote, protonated
δ-ammonium group [85]. Deprotonation to give terminal ω-amino groups
results in its insignificant interaction with Cu^{2+} in the case of Lys
and a weak apical chelation in the case of Orn according to visible
absorption and CD spectral evidence [6]. Both dap and dab complex to
Cu^{2+} as neutral ligands from pH 3 to 5 [79], and stability constant
comparisons and absorption and CD spectra all indicate chelation as a
substituted glycinate with a protonated ω-ammonium group [6, 79]. At
pH > 6 complexes of anionic ligands become predominant. For the Cu^{2+}
complexes of anionic dap visible and CD spectra indicate two nitrogen
donors in the chelate plane, whereas stability constant comparisons
suggest that at least one ligand is terdendate with apical carboxylate
chelation [6]. The two basically different kinds of Cu^{2+} complexes
typified by Lys and Orn on one hand and dap on the other come together
in the complexes of anionic dab where the results suggest several
species coexist [6]. Anionic dab serves as a terdentate ligand with

significant amounts of both the species with two nitrogens and the
species with a glycinate function in the tetragonal plane and apical
carboxylate and γ-amino groups, respectively. Histidine forms complexes analogous to those of dab (Sec. 8.2).

With Ni^{2+} weaker binding and the possibility of tris complexes
make it difficult to find conditions with excess ligand where any
complex becomes as much as 60% of all complexes present [78, 86].
The net-zero-charged ligands of Table 3 probably bind as substituted
glycinates. The anionic ligands are terdentate with the interaction
of the ε-amino group of lysine quite weak. Co^{2+} forms complexes
similar to Ni^{2+} with all four ligands [78, 87]. Zn^{2+} forms similar
complexes with dap and dab, but hydroxo complex formation occurs in
preference to chelation of the δ-amino group of ornithine and the
ε-amino group of lysine [86].

For the series of α,ω-diaminocarboxylate anions the order of
decreasing stability of Co^{2+}, Ni^{2+}, and Zn^{2+} 1:1 complexes is $dab^- >$
$dap^- > Orn^- > Lys^-$ with the stability for dab^- and dap^- equal for
Cu^{2+}. The inverse order of $dab^- > dap^-$ for Ni^{2+} has been ascribed
to a lesser stability of two five-membered chelate rings compared to
one five- and one six-membered chelate rings [78]. It has been
pointed out, however, that when allowance is made for the greater
basicity of dab^- compared to dap^- the stability of dap^- complexes
exceeds that of dab^- [87]. Therefore, the inverse order $dab^- > dap^-$
is due to the greater basicity of dab^- and two five-membered chelate
rings are actually more stable than one five- and one six-membered
ring. Though the 5,5 chelate ring system appears somewhat more
stable than the 5,6 ring system, the latter is much more stable than
the 5,7 system that occurs in Orn^-.

Strongly tetragonal Pd^{2+} chelates anionic dap and dab through
two nitrogen donors with an unbound carboxylate group, as revealed
by absorption and CD results [7, 88]. Anionic Lys is bound as a
substituted glycinate. Anionic Orn binds both through two nitrogen
donors in a seven-membered ring and as a substituted glycinate.
Thus the transition between the two kinds of binding to Pd^{2+} occurs
at anionic Orn while for Cu^{2+} it occurs at anionic dab with a

six-membered ring through two nitrogen donors. This conclusion is consistent with the greater tetragonality and proclivity of Pd^{2+} to bind nitrogen.

7.3. Homopolypeptides

The ambidentate character of α,ω-diaminocarboxylates carries over to the Cu^{2+} complexes of the homopolypeptides. Addition of Cu^{2+} to solutions containing homopolypeptides of arginine [89], lysine [90, 91], ornithine [90, 92], 2,4-diaminobutyric acid [90], and histidine [93] yields low pH < 8 complexes different from those at high pH > 12 where a biuret-type complex forms [94]. The biuret complexes of the L-polypeptides display absorption maxima at about 530 nm and a characteristic negative CD in the same region. The correspondence of these optical results to those found for the Cu^{2+} complexes of peptides [5, 19] causes us to suggest that the biuret complexes of the polypeptides are due to four deprotonated amide nitrogens located tetragonally about the Cu^{2+}. The complex of the dab polypeptide gives an exceptional visible CD pattern and a high pH 13 for full development [90]. These results suggest competition by and participation of the γ-amino groups for tetragonal Cu^{2+} sites. Both helical [90] and β- [95] secondary structures are destroyed by formation of biuret-type complexes.

The low pH complexes of the Lys and Orn homopolypeptides are fully formed at pH 7.4 after the addition of 2 equivalents base where the absorption spectrum exhibits a maximum at 650 nm [90-92]. (Substantial amounts of impurities evident in the titration curves of poly-Arg [89] cast doubt on any conclusions drawn in this case.) An absorption maximum at 650 nm indicates no more than two amino donors to Cu^{2+} with two oxygen (water) donors consistent with the titration equivalents. The low pH complex of the polypeptide of L-2,4-diamino-butyric acid is fully formed at pH 8.3 after the addition of 4 equivalents of base where the shorter absorption maximum at 580 nm has been attributed to the presence of at least one deprotonated peptide nitrogen donor atom [90]. This attribution is not necessary as four amino donors may yield a similar or even shorter wavelength: The bis-1,2-diaminopropane complex of Cu^{2+} displays an absorption maximum at 540 nm [6]. Better evidence for the coordination of one or two deprotonated peptide nitrogens is provided by the CD extremum found near 340 nm [90], which is characteristic of deprotonated peptide nitrogen coordination to Cu^{2+} [96]. Poly-L-histidine forms a Cu^{2+} complex that is fully formed by pH 4.5 after the addition of 4 equivalents of base and exhibits an absorption maximum near 540 nm [93]. The authors suggest that one deprotonated amide and three imidazole nitrogens are bound to tetragonal Cu^{2+}. The short absorption maximum would be more consistent with two amide and two imidazole nitrogens. In the low pH complexes of both of the last two polypeptides, the net positive visible CD spectra are relatively unique and resemble those found in a variety of Cu^{2+} complexes of the corresponding L-amino acids [6]. Thus in the Cu^{2+} complexes of these homopolypeptides we see the favoring at high pH of the biuret complex with four deprotonated amide nitrogens about each Cu^{2+}, and at low pH of increasingly exclusive side chain involvement as the side chain becomes longer. At low pH, only terminal amino nitrogens coordinate in the lysine and ornithine cases whereas satisfactory chelate rings involving both side chain and deprotonated peptide nitrogens evidently close in the 2,4-diaminobutyric acid and histidine polypeptide complexes.

8. HISTIDINE

8.1. The Ligand

There is an ambiguity in the numbering system of the histidine imidazole ring. Organic chemistry nomenclature prescribes that the pyrrole nitrogen is numbered 1 and the pyridine nitrogen numbered 3. The question then arises as to where on the imidazole ring the alanyl substituent is to be placed, at position 4 or 5. Much of the biochemical literature describes histidine as a 5-substituted imidazole, yet the simplest description is that of 4-substitution.

$$\text{NH}_3{}^+\text{-CH-COO}^-$$

A tautomeric equilibrium occurs between the two ring nitrogens for
the single hydrogen in a neutral imidazole ring. It has now been
established by both ^{13}C [97] and ^{15}N [98] nmr spectroscopy that the
depicted N(1)-H tautomer predominates by about 4:1 over the N(3)-H
tautomer. Therefore, drawing the structure according to the organic
chemists' conventions automatically results in depicting the favored
tautomer. An alternative designation used by some biochemists is
based on the nitrogen distance from the substituted carbon with N(3)
in the picture labeled π (for pro) and N(1) labeled τ (for tele).

The preference of histidine, acetylhistidine, and Gly-His for
an anticonformation of the imidazole ring and the carboxylate group
about the C_α-C_β bond as suggested by analysis of vicinal coupling
constants in ^{1}H nmr [99] has been supported by analysis of ^{1}H-^{13}COO$^-$
vicinal couplings in ^{13}C nmr [100]. Rotation about this bond is easy
and rapid so that alternative conformations suitable for metal ion
binding are readily adopted.

The amino acid histidine in its dicationic form contains four
acidic protons. As the pH is increased the successive pK_a values for
proton removal are carboxylic acid, 1.9; imidazolium, 6.1; ammonium,
9.1; and imidazole, 14.4 [101]. From a comprehensive histidine deprotonation scheme where 12 microconstants were evaluated, it may be
shown that in neutral solutions the predominant net-zero-charged
species with an ionized carboxylate group possesses a protonated
ammonium group by 50:1 over the species with a protonated imidazolium
group [102]. This ratio exceeds that for any of the α,ω-diamino-
carboxylates discussed in Sec. 7. The first imidazole deprotonation
with pK_a 6.1 is from the pyridine nitrogen whereas the ionization
with pK_a 14.4 is from the pyrrole nitrogen. It has been argued
strongly that metal ion binding on the aromatic imidazole ring

occurs only at a pyridine nitrogen, as the energy required for binding to occur at a pyrrole nitrogen is prohibitive [103]. For metal ion binding to occur at N(1) in the above structure, the hydrogen tautomerizes to N(3) and N(1) becomes a pyridine nitrogen. Note that chelation of the amino nitrogen with the imidazole ring to form a six-membered ring is with N(3), which is already in the appropriate pyridine form for metal ion binding in the depicted structure. As indicated in Table 1, we anticipate some analogies between dab and His as ligands.

8.2. Ligand Complexes

Many features of histidine complexes have been covered in a comprehensive review of the interactions of histidine and other imidazole derivatives with transition metal ions in chemical and biological systems [101]. There seems little need for substantial alteration of that review written 5 years ago because most subsequent articles provide support for the principles developed therein. In the short space allotted to histidine in this chapter, the emphasis is on issues that have been of most interest in the past 5 years.

As indicated by the comparisons possible in Table 2, the first histidine anion bound to a 3d transition metal ion is certainly terdentate as is the second for Co^{2+} and Ni^{2+} and, for the greater fraction probably for Zn^{2+} and Cu^{2+} as well. A recent crystal structure of D,L-histidine with Co^{3+} shows two terdentate ligands with an all cis arrangement of donor atoms [104].

The most uncertainty has concerned binding to the tetragonal Cu^{2+} ion where to coordinate all three donor groups to a single metal ion one of three donor groups must take up an axial position. These same uncertainties attended dab$^-$ in Sec. 7.2. In solutions containing some excess histidine, the main species present from pH 4 to pH 5.5 contains one anionic and one neutral histidine (species distribution curves appear in Ref. [105]). The anionic histidine is bound as a terdentate ligand with an apical carboxylate, and the neutral histidine predominantly as a substituted glycinate in the tetragonal

plane with a protonated imidazole ring. Stability constant comparisons and optical studies have been employed to reach the conclusion
that a neutral histidine binds to Cu^{2+} as just described [102]. From
pH 5.5 to about pH 11 there are about each Cu^{2+} two anionic histidines,
both of which are probably mainly terdentate. Arguments have been
presented to consider this 2:1 complex to consist of two terdentate
L-histidine anions with trans amino groups [102]. The tetragonal
plane then consists of a substituted glycinate from one histidine
and imidazole and amino nitrogens from the other. An imidazole
nitrogen from the first histidine occupies one axial position and a
carboxylate oxygen from the second histidine the other. This structure is a dynamic one with Jahn-Teller distortions carrying first one
set of trans imidazole-carboxylate groups into axial positions and
then the other. The complex is also in equilibrium with other five-
and six-coordinate structures.

The unreliability of using selective broadening of resonances
in nmr spectroscopy as a probe of paramagnetic ion binding sites has
received recent emphasis. For neither Cu^{2+} nor Mn^{2+} are ^1H or ^{13}C
resonances in imidazole primarily dipolar determined so that distances are not directly obtainable and the broadening criterion should
be used for small molecules only with discretion [38, 39, 106, 107,
and see short discussion in Sec. 7 of Chap. 2 of Vol. 8]. Detailed
studies have been made of ^{13}C relaxation in nmr spectroscopy of
histidine in the presence of the paramagnetic ions Co^{2+} [108], Mn^{2+}
[109], and Ni^{2+} [110]. ^1H nmr spectra of Co^{2+} complexes of histidine and several amino-substituted derivatives provide useful structural information [111].

8.3. Peptide Complexes

The histidyl residue provides an anchor for metal ion promoted amide
hydrogen deprotonations in peptides [112]. Of the small histidyl
containing peptides the greatest attention recently has been given
to the Cu^{2+} complexes of GlyGly-L-His which serve as a model for the

carrying of plasma exchangeable Cu^{2+} by the amino terminal AspAlaHis
residues of human plasma albumin [112-119]. As detailed in these
references and discussed previously [112], in the main species present
in equimolar neutral solutions, Cu^{2+} is chelated about the tetragonal
plane by four nitrogen donors: the amino terminal nitrogen, two deprotonated peptide nitrogens, and N(3) of the imidazole ring. There is
strong evidence that the corresponding structure occurs for the Cu^{2+}
complex of albumin [112].

Complexes of other histidyl peptides and metal ions have also
been studied [120, 121]. In equimolar, neutral solutions HisGly chelates as a substituted GlyGly with an amino, a deprotonated amide, and
a carboxylate oxygen donor atom about one Cu^{2+} with an imidazole ring
nitrogen coordinated to the fourth tetragonal position about a second
Cu^{2+} [122, 123]. This complex resembles the dimeric Cu^{2+}-carnosine
(β-alanylhistidine) complex, which has been extensively discussed
[112].

The crystal structure of a binuclear Cu^{2+} dimer of cyclo-L-His-
L-His exhibits unusual features [124]. One Cu^{2+} is five coordinate
with donor atoms from a water plus an imidazole nitrogen and an amide
carbonyl oxygen from each of the two ligands in the dimer. The other
Cu^{2+} adopts a flattened tetrahedral geometry with donor atoms from an
imidazole nitrogen and a deprotonated amide nitrogen from each of the
two ligands. Of the two necessarily cis amide bonds in each ligand,
one is not involved in coordination and the other is doubly coordinated, the carbonyl oxygen to one Cu^{2+} and the deprotonated amide
nitrogen to the other.

8.4. Pyrrole Hydrogen Deprotonations
     in Imidazole Rings

Deprotonation of the hydrogen at the imidazole ring pyrrole nitrogen
by metal ions may occur by substitution or by promotion of across
the ring ionization [125]. In both cases a metal ion is required
at the pyridine nitrogen. Substitution of the pyrrole hydrogen by

a metal ion in the known examples occurs at pH < 10. In equimolar
neutral solutions Gly-L-His serves as a terdentate ligand with amino,
deprotonated amide, and pyridine imidazole nitrogen donors to yield
complexes of Ni^{2+}, Cu^{2+}, and Pd^{2+} of net zero charge. All three
metal ion complexes undergo a concentration-dependent deprotonation
near pH 9.6, and changes in absorption spectra are indicative of
water replacement by a nitrogen donor in the fourth tetragonal position [126]. This transformation is accommodated by substitution of
the pyrrole hydrogen on one complex by the fourth coordination position about the metal ion on another to form a polymer, probably a
tetramer. The resulting complex contains an anionic imidazole ring
that bridges two metal ions, one at each nitrogen. The complexes
described present relatively rare examples of soluble bridging
imidazolate rings; examples with imidazole itself are known as
solids [127]. Interest in bridging imidazolate rings has heightened
since the discovery that His-61 in the enzyme superoxide dismutase
bridges a Cu^{2+} and a Zn^{2+} through its imidazole ring [128]. Efforts
to create and study more examples of bridging imidazolates are under-
way [129, 130].

To be contrasted with pyrrole hydrogen deprotonation by substitution is across the ring promotion of pyrrole hydrogen ionization by
a metal ion bound at the pyridine nitrogen. In the GlyHis examples
above and in the Cu^{2+} complex of GlyHisGly of similar structure, substitution of the pyrrole hydrogen occurs near pH 9.6 [131]. In the
GlyGlyHis complex, however, Cu^{2+} is chelated by an amino, two deprotonated amides, and the histidine N(3) nitrogen so that all four
tetragonal coordination positions are occupied by nitrogen donor
atoms. In this complex the presence of a metal ion at N(3) promotes
across the ring ionization, and it occurs at a higher pH (~10.7).

Occurring with pK_a ~14.4 in the unbound imidazole and histidine
ligands, the metal ion promoted across ring pyrrole hydrogen ionization may take place as low as pH 10 for imidazole rings coordinated
to 3d and divalent 4d transition metal ions [125]. Since this sum-
mary statement was made, the value for a cobalamin complexed imidazole

has been revised downward to pK_a 9.6 [132]. The low value may be attributed to the high degree of unsaturation about Co^{3+}. In support of this view, in a variety of Cu^{2+} complexes of His or histamine, pyrrole nitrogen promoted ionizations occur with pK_a 10 to 13, with mixed bipyridyl complexes at the lower end and mixed tiron complexes at the upper end of the range [133]. The imidazole complex of 5d CH_3Hg^+ exhibits pK_a = 9.6 [134]. Promoted pyrrole nitrogen ionizations do not occur in Ni^{2+} complexes at pH < 12 [133, 135]. With Co^{2+} the ionizations are associated with a change from hexacoordinate to tetrahedral stereochemistry and are estimated to occur from pH 12 to 13 [135].

It has been suggested that the activity related deprotonation occurring near pH 7 to 8 in the zinc enzyme carbonic anhydrase is due to a promoted pyrrole nitrogen ionization of one of the three side chain imidazole rings bound to Zn^{2+}. There is, however, no precedent for Zn^{2+}-promoted pyrrole nitrogen ionization in imidazole or histidine occurring at pH < 13 [136]. Due to the absence of a chemical shift from pH 10 to 13 in the ^1H nmr of the H(2) and H(5) protons on the imidazole ring of histidine in 2.5:1 solutions with Zn^{2+}, an alkali consuming process occurring at pH 10 has been assigned to hydrolysis of bound water [133]. The acidity of bound water is appreciably enhanced on binding of imidazole rings to metal ions including Zn^{2+} [137]. In the tris imidazole complex of Zn^{2+} in 3 M $NaClO_4$, hydrolysis of bound water occurs with pK_a = 8.0 [137]. That the magnitude of the third stability constant for imidazole complexation is more than eight times greater than the first suggests that a change to tetrahedral stereochemistry has occurred. Ionization of metal ion-bound water occurs more easily in complexes of lower coordination number. Since the tetrahedral tris imidazole complex is similar in structure to the Zn^{2+} site in carbonic anhydrase, the comparable acidity constant for Zn^{2+} bound water ionization in the model complex supports the common view that the activity-related deprotonation in the enzyme is also due to Zn^{2+}-bound water.

9. AROMATIC SIDE CHAINS

Leaving aside the aromatic imidazole side chain, there are suggestions in the literature of weak attractive interactions between at
least Cu^{2+} and the aromatic side chains of Phe, Tyr, and Trp.
Neither the phenolic group of Tyr nor the indole nitrogen of Trp is
involved directly in these interactions. Examination in Table 2 of
the basicity-adjusted stability constants in parentheses for Phe⁻
and the four metal ions reveals only small increments compared to
Ala⁻. However, the difference $\log K_1$ - $\log K_2$ is consistently less
for Phe⁻ and the $\log K_2$ value for Phe⁻ equals or even exceeds that
for Ala⁻ of greater basicity. This relatively large $\log K_2$ value
for Phe⁻ complexes has been noted for Cu^{2+} [10, 138, 139] and other
3d metal ions [139]; the calorimetric resolution of the difference
with Ala⁻ is not definite [138, 139]. Though their values are less
reliable than those for Phe⁻ (and uncertain for Tyr⁻ because of
competitive ammonium and phenolic deprotonations [81, 82]), stability
constants for complexes of Tyr⁻ and Trp⁻ also show reduced $\log K_1$ -
$\log K_2$ differences [27]. Though the $\log K_1$ values are unaffected,
methylation of the phenolic oxygen of Tyr and the indole nitrogen
of Trp is reported to result in the reduction of the $\log K_2$ values
for Cu^{2+} complexes by more than three log units compared to the
parent amino acids [140].

The unusual behavior of aromatic amino acid complexes of Cu^{2+}
is also displayed in the visible CD spectra. The bis Cu^{2+} complexes
of Phe⁻, Tyr⁻, and Tyr^{2-} (with ionized uncoordinated phenolate group)
all exhibit nearly the same CD magnitudes. Instead of being half
this magnitude, the mixed 1:1:1 complexes with optically inactive
Gly⁻ are augmented by 30 to 80% over the expected value [141]. Complexes with aliphatic side chains do not produce similar augmenta-
tions. These results suggest possible attractive interactions between
Cu^{2+} and the aromatic side chains.

Less than supposed van der Waals' contact distances between Cu^{2+}
and aromatic side chains have led to suggestions of weak attractive
interactions in several crystal structure determinations. In a

complex of glycyl-L-leucyl-L-tyrosine [142] and a bis complex of Tyr⁻ [143], the plane of the aromatic ring makes a small angle with the Cu^{2+} chelate plane. In neither complex is the phenolic group coordinated. In a Gly-L-Trp complex Cu^{2+} is sandwiched between two indole rings, one from the same complex and the other from a neighboring complex [144]. The indole ring nitrogen is not involved in coordination. Closer than van der Waals' contact distances are also found between Cu^{2+} and two aromatic ring positions in the bis complex of N-benzylproline [145]. On the other hand attractive interactions between Cu^{2+} and the aromatic ring are not found in a bis Phe⁻ complex [146] nor in a complex with L-valyl-L-tyrosine [147].

Any postulated weak attractive interactions between Cu^{2+} and aromatic side chains should not be confused with the interaction that gives rise to the aromatic side chain to Tb^{3+} energy transfer process that produces green luminescence from Tb^{3+} upon ultraviolet irradiation of the aromatic chromophore in many proteins [148]. This energy transfer process is a through space interaction that does not depend upon complex formation or other attractive interactions.

The prevalence and significance of weak attractive interactions between transition metal ions such as Cu^{2+} and aromatic amino acid side chains remain matters for further inquiry. Alternatively, the enhanced second stability constants may be due to attractive interactions between side chains in a bis complex.

ABBREVIATIONS

α-ab	α-aminobutyrate	dap	2,3-diaminopropanoate
Ala	alanine	Gln	glutamine
Arg	arginine	Glu	glutamate
Asn	asparagine	Gly	glycine
Asp	aspartate	His	histidine
CD	circular dichroism	Lys	lysine
Cys	cysteine	Met	methionine
dab	2,4-diaminobutanoate	nmr	nuclear magnetic resonance

Orn   ornithine   Thr   threonine

Phe   phenylalanine   Trp   tryptophan

Ser   serine   Tyr   tyrosine

SMC   S-methylcysteine

ADDENDUM

Stability constants have been reported for chelation of tetragonal
$enPd^{2+}$ with several amino acid ligands [149]. Since the bidentate-
substituted glycinate binding mode is also available in methionine
and S-methylcysteine, comparison with stability and acidity constant
values for the other ligands presented indicates that for the glycinate binding mode their stability constant logarithms should be about
10.8 and 10.6, respectively. That the reported values are 1.7 and
1.2 log units less [149] suggests that the results for these two
ligands should be reinterpreted. Only bidendate chelation is probable
to the strongly tetragonal Pd^{2+} and N,S chelation is known to occur
with both Met and SMC [7]. It is likely that the carboxylic acid
group deprotonates from already chelated ligands with $pK_a \simeq 3$, and
that the carboxylate group formed remains unbound. Since these sentences were written Dr. M. C. Lim has kindly communicated a reanalysis
of the titration curve of methionine with $enPd^{2+}$. The curve is fitted
well by considering an N,S-chelated ligand undergoing an unbound carboxylic acid group ionization with $pK_a = 2.77$.

    A crystal structure determination of the mixed complex of
anionic L- and D-histidine with Cu^{2+} shows an all trans arrangement
of donor groups with four nitrogens in the tetragonal plane [150].
Neither this structure nor that proposed for the unmixed complex
with anionic L-histidine in solution accounts for the electronic
absorption and CD spectra of solutions containing histidine and Cu^{2+}
[6, 102]. The structure found in the crystal and that proposed for
solutions [150] can only represent minor species in solution.

REFERENCES

1. G. L. Nelsestuen, M. Broderius, T. H. Zytkovicz, and J. B. Howard, *Biochem. Biophys. Res. Commun., 65,* 233 (1975).

2. R. K. Boggess and R. B. Martin, *J. Amer. Chem. Soc., 97,* 3076 (1975).

3. P. M. May, P. W. Linder, and D. R. Williams, *J. Chem. Soc. Dalton,* 588 (1977).

4. D. D. Perrin and R. P. Agarwal, in Metal Ions in Biological Systems (H. Sigel, ed.), Vol. 2, Chap. 4, Marcel Dekker, New York, 1973.

5. R. B. Martin in Metal Ions in Biological Systems (H. Sigel, ed.), Vol. 1, Chap. 4, Marcel Dekker, New York, 1974.

6. E. W. Wilson, Jr., M. H. Kasperian, and R. B. Martin, *J. Amer. Chem. Soc., 92,* 5365 (1970).

7. E. W. Wilson, Jr. and R. B. Martin, *Inorg. Chem., 9,* 528 (1970).

8. O. Yamauchi, Y. Nakao, and A. Nakahara, *Bull. Chem. Soc. Jap., 48,* 2572 (1975).

9. T. Sakurai, O. Yamauchi, and A. Nakahara, *Bull. Chem. Soc. Jap., 49,* 169, 1579 (1976).

10. R. B. Martin and R. Prados, *J. Inorg. Nucl. Chem., 36,* 1665 (1974).

11. L. I. Katzin and E. Gulyas, *J. Amer. Chem. Soc., 91,* 6940 (1969).

12. R. A. Haines and M. Reimer, *Inorg. Chem., 12,* 1482 (1973).

13. F. S. Stephens, R. S. Vagg, and P. A. Williams, *Acta Crystallogr., B31,* 841 (1975); *B33,* 433 (1977).

14. R. J. Flook, H. C. Freeman, C. J. Moore, and M. L. Scudder, *J. Chem. Soc. Chem. Commun.,* 753 (1973).

15. A. Gergely, I. Nagypál, and E. Farkas, *J. Inorg. Nucl. Chem., 37,* 551 (1975).

16. M-C. Lim, *J. Chem. Soc. Dalton,* 1398 (1977).

17. A. Gergely, J. Mojzes, and Z. Kassai-Bazsa, *J. Inorg. Nucl. Chem., 34,* 1277 (1972).

18. L. D. Pettit and J. L. M. Swash, *J. Chem. Soc. Dalton,* 2416 (1976).

19. J. M. Tsangaris and R. B. Martin, *J. Amer. Chem. Soc., 92,* 4255 (1970).

20. P. Grenouillet, R.-P. Martin, A. Rossi, and M. Ptak, *Biochim. Biophys. Acta, 322,* 185 (1973).

21. D. B. McCormick, H. Sigel, and L. D. Wright, *Biochim. Biophys. Acta, 184,* 318 (1969).

22. D. van der Helm and M. B. Hossain, *Acta Crystallogr., B25,* 457 (1969).

23. D. van der Helm and W. A. Franks, *Acta Crystallogr., B25,* 451 (1969).

24. D. van der Helm, A. F. Nicholas, and C. G. Fisher, *Acta Crystallogr., B26,* 1172 (1970).

25. H. C. Freeman, J. M. Guss, M. J. Healy, R-P. Martin, C. E. Nockolds, and B. Sarkar, *Chem. Commun.,* 225 (1969).

26. H. Sigel, C. F. Naumann, B. Prijs, D. B. McCormick, and M. C. Falk, *Inorg. Chem., 16,* 790 (1977).

27. A. E. Martell and R. M. Smith, Critical Stability Constants, Vols. 1 and 2, Plenum Press, New York, 1974 and 1975.

28. P. Bonvicini, A. Levi, V. Lucchini, and G. Scorrano, *J. Chem. Soc. Perkin II,* 2267 (1972).

29. P. Bonvicini, A. Levi, V. Lucchini, G. Modena, and G. Scorrano, *J. Amer. Chem. Soc., 95,* 5960 (1973).

30. D. J. Lee and R. Cameron, *J. Amer. Chem. Soc., 93,* 4724 (1971).

31. S. Ahrland, J. Chatt, and N. R. Davies, *Quart. Rev. Chem. Soc., 12,* 265 (1958).

32. R. G. Pearson (ed.), Hard and Soft Acids and Bases, Dowden, Hutchinson, and Ross, Stroudsberg, Pennsylvania, 1973.

33. D. F. S. Natusch and L. J. Porter, *J. Chem. Soc. A,* 2527 (1971); *Chem. Commun.,* 596 (1970).

34. D. B. McCormick, R. Griesser, and H. Sigel, in Metal Ions in Biological Systems (H. Sigel, ed.), Vol. 1, Chap. 6, Marcel Dekker, New York, 1974.

35. R. Griesser, M. G. Hayes, D. B. McCormick, B. Prijs, and H. Sigel, *Arch. Biochem. Biophys., 144,* 628 (1971).

36. E. Gonick, W. C. Fernelius, and B. E. Douglas, *J. Amer. Chem. Soc., 76,* 4671 (1954).

37. D. B. McCormick, H. Sigel, and L. D. Wright, *Biochim. Biophys. Acta, 184,* 318 (1969).

38. W. G. Espersen and R. B. Martin, *J. Phys. Chem., 80,* 161 (1976).

39. W. G. Espersen and R. B. Martin, *J. Amer. Chem. Soc., 98,* 40 (1976).

40. V. M. Miskowski, J. A. Thich, R. Solomon, and H. J. Schugar, *J. Amer. Chem. Soc., 98,* 8344 (1976).

41. T. E. Jones, D. B. Rorabacher, and L. A. Ochrymouycz, *J. Amer. Chem. Soc., 97,* 7485 (1975).

42. M. V. Veidis and G. J. Palenik, *Chem. Commun.*, 1277 (1969).

43. R. B. Wilson, P. de Meester, and D. J. Hodgson, *Inorg. Chem.*, *16*, 1498 (1977).

44. R. J. Flook, H. C. Freeman, C. J. Moore, and M. L. Scudder, *J. Chem. Soc. Chem. Commun.*, 753 (1973).

45. C. A. Bear and H. C. Freeman, *Acta Crystallogr.*, *B32*, 2534 (1976).

46. P. de Meester and D. J. Hodgson, *J. Amer. Chem. Soc.*, 6884 (1977).

47. C. C. Ou, V. M. Miskowski, R. A. Lalancette, J. A. Potenza, and H. J. Schugar, *Inorg. Chem.*, *15*, 3157 (1976).

48. C. A. McAuliffe, J. V. Quagliano, and L. M. Vallarino, *Inorg. Chem.*, *5*, 1996 (1966).

49. C. A. McAuliffe, *J. Chem. Soc. A*, 641 (1967).

50. S. E. Livingstone and J. D. Nolan, *Inorg. Chem.*, *7*, 1447 (1968).

51. J. Hidaka, S. Yamada, and Y. Shimura, *Chem. Lett.*, 1487 (1974).

52. P. de Meester and D. J. Hodgson, *J. Chem. Soc. Dalton*, 618 (1976).

53. V. M. Kothari and D. H. Busch, *Inorg. Chem.*, *8*, 2276 (1969).

54. L. M. Volshtein, L. F. Krylova, and M. F. Mogilevkina, *Russ. J. Inorg. Chem.*, *12*, 832 (1967).

55. R. C. Warren, J. F. McConnell, and N. C. Stephenson, *Acta Crystallogr.*, *B26*, 1402 (1970).

56. L. P. Battaglia, A. B. Corradi, C. G. Palmieri, M. Nardelli, and M. E. V. Tani, *Acta Crystallogr.*, *B29*, 762 (1973).

57. H. C. Freeman and M. L. Golomb, *Chem. Commun.*, 1523 (1970).

58. L. E. Erickson, J. W. McDonald, J. K. Howie, and R. P. Clow, *J. Amer. Chem. Soc.*, *90*, 6371 (1968).

59. P. Haake and P. C. Turley, *J. Amer. Chem. Soc.*, *89*, 4611 (1967); P. C. Turley and P. Haake, ibid., *89*, 4617 (1967).

60. V. M. Rheinberger and H. Sigel, *Naturwissenschaften*, *62*, 182 (1975).

61. A. J. Carty and N. J. Taylor, *J. Chem. Soc. Chem. Commun.*, 214 (1976).

62. M. T. Fairhurst and D. L. Rabenstein, *Inorg. Chem.*, *14*, 1413 (1975); D. L. Rabenstein, *Accts. Chem. Res.*, *11*, 100 (1978).

63. Y.-S. Wong, A. J. Carty, and P. C. Chieh, *J. Chem. Soc. Dalton*, 1157 (1977).

64. N. Mandel, G. Mandel, B. L. Trus, J. Rosenberg, G. Carlson, and R. E. Dickerson, *J. Biol. Chem.*, *252*, 4619 (1977).

65. P. M. Colman, H. C. Freeman, J. M. Guss, M. Murata, V. A. Norris, J. A. M. Ramshaw, and M. P. Venkatappa, *Nature, 272,* 319 (1978).

66. M. M. David, R. Sperling, and I. Z. Steinberg, *Biochim. Biophys. Acta, 359,* 101 (1974).

67. C. J. Hawkins and D. D. Perrin, *Inorg. Chem., 2,* 843 (1963).

68. J. F. Boas, J. R. Pilbrow, C. D. Hartzell, and T. D. Smith, *J. Chem. Soc. A,* 572 (1969).

69. E. W. Wilson, Jr. and R. B. Martin, *Arch. Biochem. Biophys., 142,* 445 (1971).

70. W. G. Jackson, A. M. Sargeson, and P. A. Tucker, *J. Chem. Soc. Chem. Commun.,* 199 (1977).

71. J. A. Thich, D. Mastropaolo, J. Potenza, and H. J. Schugar, *J. Amer. Chem. Soc., 96,* 726 (1974).

72. P. E. Riley and K. Seff, *Inorg. Chem., 11,* 2993 (1972).

73. T. Ottersen, L. G. Warner, and K. Seff, *Inorg. Chem., 13,* 1904 (1974).

74. C.-I. Brändén, *Acta Chem. Scand., 21,* 1000 (1967).

75. L. G. Warner, T. Ottersen, and K. Seff, *Inorg. Chem., 13,* 2819 (1974).

76. R. J. Gale and C. A. Winkler, *Inorg. Chim. Acta, 21,* 151 (1976).

77. R. W. Hay and P. J. Morris, *J. Chem. Soc. Perkin II,* 1021 (1972).

78. G. Brookes and L. D. Pettit, *J. Chem. Soc. Dalton,* 42 (1976).

79. A. Gergely, E. Farkas, I. Nagypál, and E. Kas, *J. Inorg. Nucl. Chem., 40,* 1709 (1978).

80. R. B. Martin, Introduction to Biophysical Chemistry, McGraw-Hill Book Co., New York, 1964, pp. 67-77.

81. J. T. Edsall, R. B. Martin, and B. R. Hollingworth, *Proc. Nat. Acad. Sci. U.S., 44,* 505 (1958).

82. R. B. Martin, J. T. Edsall, D. B. Wetlaufer, and B. R. Hollingworth, *J. Biol. Chem., 233,* 1429 (1958).

83. R. B. Martin, *J. Phys. Chem., 75,* 2657 (1971).

84. T. L. Sayer and D. L. Rabenstein, *Can. J. Chem., 54,* 3392 (1976).

85. F. S. Stephens, R. S. Vagg, and P. A. Williams, *Acta Crystallogr., B33,* 438 (1977).

86. E. Farkas, A. Gergely, and E. Kas, *J. Inorg. Nucl. Chem.,* in press.

87. M. Gold and H. K. J. Powell, *J. Chem. Soc. Dalton,* 1418 (1976).

88. T. P. Pitner, E. W. Wilson, Jr., and R. B. Martin, *Inorg. Chem., 11,* 738 (1972).

89. A. Garnier and L. Tosi, *Biopolymers, 14,* 2247 (1975); L. Tosi and A. Garnier, *Biochem. Biophys. Res. Commun., 58,* 427 (1974).

90. M. Palumbo, A. Cosani, M. Terbojevich, and E. Peggion, *Macro-molecules, 10,* 813 (1977); *J. Amer. Chem. Soc., 99,* 939 (1977).

91. A. Garnier and L. Tosi, *Biochem. Biophys. Res. Commun., 74,* 1280 (1977).

92. C. Phan, L. Tosi, and A. Garnier, *Bioinorganic Chem., 8,* 21 (1978).

93. A. Levitzki, I. Pecht, and A. Berger, *J. Amer. Chem. Soc., 94,* 6844 (1972).

94. A. S. Brill, R. B. Martin, and R. J. P. Williams, in Electronic Aspects of Biochemistry (B. Pullman, ed.), Academic Press, New York, 1964, pp. 519-557.

95. M. Palumbo, A. Cosani, M. Terbojevich, and E. Peggion, *Biopoly-mers, 17,* 243 (1978).

96. J. M. Tsangaris, J. W. Chang, and R. B. Martin, *J. Amer. Chem. Soc., 91,* 726 (1969).

97. W. F. Reynolds, I. R. Peat, M. H. Freedman, and J. R. Lyera, Jr., *J. Amer. Chem. Soc., 95,* 328 (1973).

98. F. Blomberg, W. Maurer, and H. Ruterjans, *J. Amer. Chem. Soc., 99,* 8149 (1977).

99. R. B. Martin and R. Mathur, *J. Amer. Chem. Soc., 87,* 1065 (1965).

100. W. G. Espersen and R. B. Martin, *J. Phys. Chem., 80,* 741 (1976).

101. R. J. Sundberg and R. B. Martin, *Chem. Rev., 74,* 471 (1974).

102. Ref. 101, Sec. IV.

103. Ref. 101, Sec. I.

104. N. Thorup, *Acta Chem. Scand., A31,* 203 (1977).

105. T. P. A. Kruck and B. Sarkar, *Can. J. Chem., 51,* 3549 (1973).

106. W. G. Espersen, W. C. Hutton, S. T. Chow, and R. B. Martin, *J. Amer. Chem. Soc., 96,* 8111 (1974).

107. R. E. Wasylishen and M. R. Graham, *Can. J. Chem., 54,* 617 (1976).

108. S. Kitagawa, K. Yoshikawa, and I. Morishima, *J. Phys. Chem., 82,* 89 (1978).

109. J. J. Led and D. M. Grant, *J. Amer. Chem. Soc., 97,* 6962 (1975).

110. J. J. Led and D. M. Grant, *J. Amer. Chem. Soc., 99,* 5845 (1977).

111. J. H. Ritsma, *J. Inorg. Nucl. Chem., 38,* 907 (1976).

112. Ref. 101, Sec. VI.

113. S. Lau, T. P. A. Kruck, and B. Sarkar, *J. Biol. Chem., 249,* 5878 (1974).

114. T. P. A. Kruck, S. Lau, and B. Sarkar, *Can. J. Chem.*, *54*, 1300 (1976).

115. N. Camerman, A. Camerman, and B. Sarkar, *Can. J. Chem.*, *54*, 1309 (1976).

116. T. P. A. Kruck and B. Sarkar, *Inorg. Chem.*, *14*, 2383 (1975).

117. S. Lau and B. Sarkar, *Can. J. Chem.*, *53*, 710 (1975).

118. R. P. Agarwal and D. D. Perrin, *J. Chem. Soc. Dalton*, 53 (1977).

119. J. C. Cooper, L. F. Wong, D. L. Venezky, and D. W. Margerum, *J. Amer. Chem. Soc.*, *96*, 7560 (1974).

120. R. P. Agarwal and D. D. Perrin, *J. Chem. Soc. Dalton*, 268, 1045 (1975); 89 (1976).

121. R. Österberg and B. Sjöberg, *J. Inorg. Nucl. Chem.*, *37*, 815 (1975).

122. H. Aiba, A. Yokoyama, and H. Tanaka, *Bull. Chem. Soc. Jap.*, *47*, 136 (1974).

123. R. K. Boggess and R. B. Martin, *J. Inorg. Nucl. Chem.*, *37*, 1097 (1975).

124. Y. Kojima, K. Hirotsu, and K. Matsumoto, *Bull. Chem. Soc. Jap.*, *50*, 3222 (1977).

125. Ref. 101, Sec. IX.

126. P. J. Morris and R. B. Martin, *J. Inorg. Nucl. Chem.*, *33*, 2913 (1971).

127. Ref. 101, Sec. III.

128. J. S. Richardson, K. A. Thomas, B. H. Rubin, and D. C. Richardson, *Proc. Nat. Acad. Sci. U.S.*, *72*, 1349 (1975).

129. G. Kolks, C. R. Frihart, H. N. Rabinowitz, and S. J. Lippard, *J. Amer. Chem. Soc.*, *98*, 5720 (1976).

130. G. Kolks and S. J. Lippard, *J. Amer. Chem. Soc.*, *99*, 5804 (1977).

131. H. Aiba, A. Yokoyama, and H. Tanaka, *Bull. Chem. Soc. Jap.*, *47*, 1437 (1974).

132. W. J. Eilbeck and M. S. West, *J. Chem. Soc. Dalton*, 274 (1976).

133. I. Sóvágó, T. Kiss, and A. Gergely, *J. Chem. Soc. Dalton*, *964* (1978).

134. C. A. Evans, D. L. Rabenstein, G. Geier, and I. W. Erni, *J. Amer. Chem. Soc.*, *99*, 8106 (1977).

135. P. J. Morris and R. B. Martin, *J. Amer. Chem. Soc.*, *92*, 1543 (1970).

136. R. B. Martin, *Proc. Nat. Acad. Sci. U.S.*, *71*, 4346 (1974).

137. W. Forsling, *Acta Chem. Scand.*, *A31*, 759 (1977).

138. R. M. Izatt, J. W. Wrathall, and K. P. Anderson, *J. Phys. Chem.*, *65*, 1914 (1961).

139. A. Gergely, I. Nagypál, and B. Király, *Acta Chim. Acad. Sci. Hung.*, *68*, 285 (1971).

140. O. A. Weber, *J. Inorg. Nucl. Chem.*, *36*, 1341 (1974).

141. E. W. Wilson, Jr. and R. B. Martin, *Inorg. Chem.*, *10*, 1197 (1971).

142. W. A. Franks and D. van der Helm, *Acta Crystallogr.*, *B27*, 1299 (1970); D. van der Helm and W. A. Franks, *J. Amer. Chem. Soc.*, *90*, 5627 (1968).

143. D. van der Helm and C. E. Tatsch, *Acta Crystallogr.*, *B28*, 2307 (1972).

144. M. B. Hursthouse, S. A. A. Jayaweera, H. Milburn, and A. Quick, *J. Chem. Soc. Dalton*, 2569 (1975).

145. G. G. Aleksandrov, Y. T. Struchkov, A. A. Kurganov, S. V. Rogozhin, and V. A. Davankov, *J. Chem. Soc. Chem. Commun.*, 1328 (1972).

146. D. van der Helm, M. B. Lawson, and E. I. Enwall, *Acta Crystallogr.*, *B27*, 2411 (1971).

147. V. Amirthalingam and K. V. Muralidharan, *Acta Crystallogr.*, *B32*, 3153 (1976).

148. H. G. Brittain, F. S. Richardson, and R. B. Martin, *J. Amer. Chem. Soc.*, *98*, 8255 (1976).

149. M. C. Lim, *J. Chem. Soc. Dalton*, *726* (1978).

150. N. Camerman, J. K. Fawcett, T. P. A. Kruck, B. Sarkar, and A. Camerman, *J. Amer. Chem. Soc.*, *100*, 2690 (1978).

Chapter 2

METAL COMPLEXES OF ASPARTIC ACID AND GLUTAMIC ACID

Christopher A. Evans, Roger Guevremont,
and Dallas L. Rabenstein
Department of Chemistry
The University of Alberta
Edmonton, Alberta, Canada

1. INTRODUCTION

1.1. General Remarks

Aspartic acid (I) and glutamic acid (II) both have three binding
sites for protons or metal ions. Chelation of the amino and

$$^+NH_3-CH-CO_2^- \qquad\qquad ^+NH_3-CH-CO_2^-$$
$$\qquad CH_2 \qquad\qquad\qquad\qquad CH_2$$
$$\qquad CO_2H \qquad\qquad\qquad\qquad CH_2$$
$$\qquad\qquad\qquad\qquad\qquad\qquad\qquad CO_2H$$

$$\text{I} \qquad\qquad\qquad\qquad\qquad \text{II}$$

α-carboxylate groups occurs readily, and much of the study of metal
binding has been devoted to attempts to deduce whether the side
chain carboxylates are also coordinated to the metal ion. Three
chelate rings are formed when these ligands are tridentate, one
glycine-like five-membered ring and rather unusual six-, seven-, or
eight-membered rings. The majority of convincing reports suggest
that aspartate will usually be tridentate whereas no well-established
case of mononuclear complexes with tridentate glutamate has been
reported. Although to date crystal structures of glutamate complexes
have invariably shown bridging of two or even three metal ions, the
possibility of polymerization has been little considered in solution
studies. A few cases of binding or chelation not involving the amine
group have been reported.

Most studies have been by the pH metric proton competition
method, which gives only partial structural information because the
carboxylate groups, in particular the α-carboxylate, are often
ionized at those pH's where complexation occurs. Stability constants

of quite different magnitudes have been reported by different groups
of investigators, and the danger of overinterpretation of data is
always present. Also, the use of potentiometric titrations to determine stability constants sometimes fails to detect polynuclear complexes which, if they exist, become important only at higher concentrations. Deduction of structures from the magnitudes of equilibrium
stability constants is risky, but at present little auxiliary information is available. Several crystal structures have been reported
but many of the more interesting structures, particularly those of
mixed complexes, await determination. ^{1}H nmr and some ^{13}C nmr
studies have been useful, and vibrational spectroscopy could be
exploited more frequently.

In this chapter we review the coordination chemistry of the
title compounds. Not all work could be covered in the space permitted; we hope that what has been included will give an understanding of the ways in which the side chains of these ligands effect
properties in them different from the bifunctional amino acids. The
corresponding amides, asparagine and glutamine, are not specifically
covered (although occasional mention must inevitably be made); nor
are peptides containing these amino acids as residues.

1.2. Occurrence

Aspartic acid was first prepared from the acid hydrolyzate of asparagine, isolated from asparagus, in 1806 [1]. Confusion over its composition was resolved in part by the work of Pasteur [2] and Kolbe
[3], 50 years later. The copper and calcium salts were used by
Ritthausen [4, 5] to aid crystallization from the mother liquors.
It is found free in fruit [6] and vegetable juices, plasma [7], and
is a constituent of proteins [5, 6]. Glutamic acid was isolated by
Ritthausen [8-11] from the gluten of wheat flour, and salts of barium,
copper, and silver were prepared. It crystallized well (solubility
is often a problem in studying this amino acid) and has been found
to be similarly widely distributed.

1.3. Aspects of Coordination

In most cases, aspartate is tridentate to a single metal ion, whereas glutamate is bidentate with the side chain extended. Here we list some of the possible structure types for these situations, and the implications thereof. The conformations of five- and six-membered chelate rings have been discussed [12, 13]; generally, glycinate rings are closely planar although coordination of the side chain carboxylate of aspartate leads to nonplanarity of this ring [13].

1.3.1. Octahedral Coordination

(a) *Tris complexes*. Tris complexes have been characterized most fully for the substitution-inert cobalt(III) ion, and these involve coordination through the glycinate ring. Two geometrical isomers, one with all three nitrogen atoms in mutually cis positions and the other with one pair of nitrogen atoms in trans positions, are possible. In addition, the distribution of the rings is chiral and, because the amino acids are themselves optically active, each configurational isomer is a diastereomer [12, 14]. A balance of interligand interactions, both repulsive and attractive, leads to considerable differences in stability for the four diastereomers of cobalt(III) complexes, and a complete description of the formation of tris complexes of labile metals would require, inter alia, a technique to obtain geometrical and diastereomer ratios of the product.

(b) *Bis-bidentate complexes*. Bis(glutamato) complexes are of the bis-bidentate type. There are five geometrical isomers [12], three of which have chiral configurations. In the case of metals with a tetragonal distortion, it seems likely that the glycinate ring will coordinate in the equatorial plane (see Sec. 4.4).

(c) *Bis-tridentate complexes*. Because the binding of the side chain carboxylate is limited to a single position relative to the glycinate ring, only three geometrical isomers (a, b, c, in Fig. 1) are possible for a bis(tridentate) complex of S-aspartate.

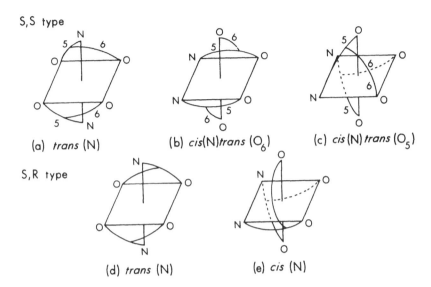

FIG. 1. Geometrical isomers for bis(aspartato)metallate with
tridentate aspartate. (Reprinted, slightly adapted, with permission from The Bulletin of the Chemical Society of Japan, *50*, 2288
(1977), by H. Oki and Y. Takahashi. Copyright by the Chemical
Society of Japan.)

For the meso complex, with ligands of opposite optical form, two
geometrical isomers (d, e; Fig. 1) are possible.

1.3.2. Square Planar Coordination

Two geometrical isomers, cis and trans, are possible. Generally,
the side chains will be directed away from the metal ion, but if
there is some apical coordination [as for Cu(II), for example],
then the cis isomer, which places the side chains on opposite sides
of the equatorial plane, should be more stable than the trans. The
cis isomer of the meso complex places both side chains on the same
side of the plane; if cis and trans geometrical isomers, per se,
are of unequal stability these systems should show stereoselectivity.

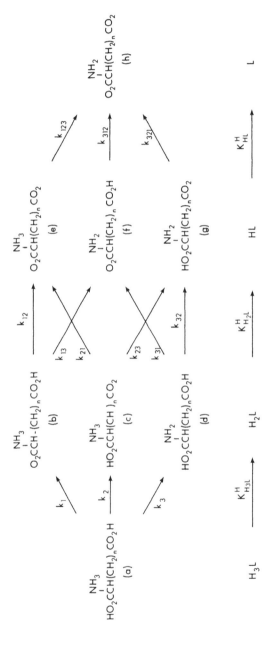

FIG. 2.   Microscopic and macroscopic ionization scheme for aspartic ($n = 1$) and glutamic ($n = 2$) acids.   Compare with Table 1.

2. ACID-BASE CHEMISTRY

2.1. Acid Dissociation Constants

The proton dissociation scheme for fully protonated aspartic or glutamic acid is shown in Fig. 2. Determination of all the constants in this scheme is difficult and to date no fully satisfactory set of constants has been reported; however, to a good approximation the ionization is sequential with protons being titrated from the α-carboxylate, the side chain carboxylate, and finally the ammonium group on the addition of one, two, and three moles of base, respectively. This corresponds to the route via k_1, k_{12}, k_{123} in Fig. 2, and the macroscopic constants $K^H_{H_3L}$, $K^H_{H_2L}$, and K^H_{HL} have values not too different from the values listed in Table 1 for these microscopic constants.

2.2. Microscopic Constants

The only practicable approach to evaluating the microconstants of these acids has been to prepare the three esters and measure their ionization constants. Implicit in this procedure is the assumption

TABLE 1

Microscopic Dissociation Constants of Glutamic Acid[a]

	$C_\alpha O_2 H$		$C_\delta O_2 H$		NH_3
pk_1	2.15	pk_2	3.85	pk_3	7.04
pk_{21}	2.62	pk_{12}	4.32	pk_{13}	9.19
pk_{31}	4.30	pk_{32}	4.65	pk_{23}	7.84
pk_{231}	4.74	pk_{132}	5.09	pk_{123}	9.96

Note: Using information in Ref. 19, we calculate, for aspartic acid, $pk_1 = 2.0$, $pk_{21} = 2.4$, $pk_2 = 3.2$, $pk_{12} = 3.7$, $pk_3 = 6.5$.

[a]From Ref. 15. The constants are identified in Fig. 2.

that the ester group has the same effect on the ionization constant
of other groups in the molecule as a carboxyl group. Microconstants
have been estimated in this way for glutamic acid [15-18] and, less
fully, for aspartic acid [19]. The values of the constants are
listed in Table 1; those for aspartic acid should be treated with
caution since the tautomer ratios given by Edsall and Blanchard [19]
are derived from empirical relationships, not experiment.

Microconstants can sometimes be derived from spectroscopic
measurements in situations where sensible proportions of the minor
tautomers are formed [20, 21] but most likely such methods would not
be successful for these molecules in view of the rather large proportion of the molecules which ionize through the major pathway [21]
and the small number of bonds between the two carboxyl groups [22].

2.3. Macroscopic Constants

The macroscopic constants have been determined on innumerable occa-
sions and may be found in almost any publication dealing with the
evaluation of metal-binding constants. Values of $pK_{H_3L}^H = 1.9$,
$pK_{H_2L}^H = 3.7$, and $pK_{HL}^H = 9.6$ are generally found for aspartic acid
at 25°C and ionic strength ~0.1 M; the corresponding values for
glutamic acid are 2.1, 4.1, and 9.5. The thermodynamics of dissociation have been reported [23-28].

Llopis and Ordonez [23] conclude that there is a close association between the two carboxylate groups and the ammonium group of
monoprotonated aspartic acid, but that for glutamic acid only the
α-carboxylate associates with the ammonium group. Interestingly,
this binding of a proton is similar to the coordination of metal
ions. Studies of the populations of the three rotational isomers
of aspartic acid [29-33] provide evidence that, in the monoprotonated
form, 90% of the aspartic acid exists as those two isomers in which
there is close association of the side chain and the ammonium group,
with most of the molecules having the two carboxylates in trans
positions [32-34]. In an elegant extension of this work, the

dissociation constants of the ammonium group in each rotational
isomer of aspartic acid have been calculated [34]. The crystal
structures of S-aspartic acid [35], S,R-aspartic acid [36], and
S-glutamic acid [37] show the α-carboxylate group ionized and the
side chain carboxylate protonated, as expected.

3. COMPLEXES WITH MAIN GROUP METAL IONS

3.1. Alkali and Alkaline Earth Metal Ions

Bonding to these metal ions is usually ionic and characterized by
lack of strong geometrical requirements. Considerable attention has
been devoted to the calcium-binding proteins, many of which have
very high ratios of aspartyl and glutamyl residues, and the crystal
structures of several such proteins show metal binding by the side
chain carboxylates of these amino acids [38]. Nevertheless, reports
of studies of binding of these metal ions by aspartic or glutamic
acids themselves are sparse.

3.1.1. *Sodium*

The optical rotation of a solution of sodium molybdate and aspartic
acid, at a pH where the Mo(VI) is entirely as MoO_4^{2-}, is different
from that of aspartic acid alone at that pH. This was attributed to
the formation of a sodium ion-aspartic acid complex [39]. However,
the ^{23}Na nmr spin lattice relaxation times of sodium chloride-
aspartic acid solutions show that any interaction between Na^+ and
the various solution forms of aspartic acid is slight [40], and it
may be that the optical rotation results can be better interpreted
as a Pfeiffer effect.

3.1.2. *Magnesium, Calcium, Strontium, and Barium*

Stability constants for aspartate and glutamate complexes of all of
these metal ions have been reported [41-46]. Lumb and Martell [41]
demonstrate a closely linear relationship between log K_{ML}^M and the

TABLE 2

Selected Equilibrium Stability Constants

A. Aspartic Acid Complexes

Metal	Method	Conditions	log K_{ML}^{M}	log $K_{ML_2}^{ML}$	log β(pqr)	log β(pqr)	Ref.
Mg(II)	pot.	25, 0.1(KCl)	2.43				41
Ca(II)	pot.	25, 0.1(KCl)	1.60				41
	pot.	25, 0.7(KCl)	1.53				46
Sr(II)	pot.	25, 0.1(KCl)	1.48				41
Ba(II)	pot.	25, 0.1(KCl)	1.14				41
Ra(II)	ix	25, 0.16(NaCl)	0.86				51
Zn(II)	pot.	37, 0.15(NaClO$_4$)	6.01	4.09	11.88(111)		54
	pot.	25, 0.3(NaClO$_4$)	5.66	4.48	10.78(111)	-10.41(1-22)	55
	pot.	25, 0.2(KCl)	5.73	4.47			59
Cd(II)	pot.	25, 0.1(KNO$_3$)	4.7	3.4			58
	pot., Cd.	25, 0.1(KNO$_3$)	4.85	3.35			58
In(III)	pot.	25, 0.1(NaClO$_4$)			12.96(111)	22.66(122)	70
Tl(I)	pot.	25, 0.1(LiClO$_4$)	2.50				71
Pb(II)	pol.	25, 0.3(NaClO$_4$)	6.0				73
Co(II)	pot.	25, 0.1(KNO$_3$)	5.94	4.29			104
Ni(II)	pot.	25, 0.2(KCl)	7.14	5.29	11.19(111)		108
	pot.	25, 0.1(NaClO$_4$)	7.16	5.24			104
Cu(II)	pot.	25, 0.2(KCl)	8.80	6.96	12.48(111)		27
	pot.	25, 0.1(KNO$_3$)	8.94	6.95	12.59(111)		121
	pot.	25, 0.1(KNO$_3$)	9.08	7.17	12.82(111) 21.21(112)	25.15(122)	122
Fe(III)	pot.	20, 1.0(NaClO$_4$)	11.4				140

Fe(II)	pot.	20, 1.0(KCl)	4.34				141
Mn(II)	pot.	25, 0.1(KCl)	3.74				143
VO(IV)	pot.	25, 0.1(KNO_3)	8.39	6.05			144
Pd(II)	pot.	30, 0.1(KNO_3)	10.55	7.7			150

B. Glutamic Acid Complexes

Mg(II)	pot.	25, 0.1(KCl)	1.9				41
Ca(II)	sol.	25	2.06		11.12(111)		45
	pot.	25, 0.1(KCl)	1.43				41
	pot.	25, 0.7(KCl)	1.20				46
Sr(II)	pot.	25, 0.1(KCl)	1.37				41
Ba(II)	pot.	25, 0.1(KCl)	1.28				41
Zn(II)	pot.	25, 0.2(KCl)	4.69	3.86			59
Cd(II)	pot.	25, 0.1(KNO_3)	4.0	3.1			58
	pot., Cd.	25, 0.1(KNO_3)	4.2	2.85			58
Ga(III)	pot.	25, 3.0($NaClO_4$)	11.30		14.19(111)	16.73(121)	68
Tl(I)	pot.	25, 0.1($LiClO_4$)	2.35				71
Pb(II)	pot.	25, 0.3($NaClO_4$)	4.70				74
Co(II)	pot.	25, 0.15(KNO_3)	4.56	3.30			104
Ni(II)	pot.	25, 0.1($NaClO_4$)	5.62	4.20			104
Cu(II)	pot.	25, 0.2(KCl)	8.27	6.47	12.39(111)		27
	pot.	25, 0.1($NaClO_4$)	8.39	6.54	12.49(111)		121
	pot.	25, 0.1(KNO_3)	8.55	6.67	12.73(111) 20.57(121)	25.18(122)	122
Fe(III)	pot.	20, 1.0($NaClO_4$)	12.1				140
Fe(II)	pot.	20, 0.1(KCl)	3.52				141
Mn(II)	pot.	20, 0.02(K_2SO_4)	3.3				52
Pd(II)	pot.	25, 0.1($NaClO_4$)	10.38	7.46			151

51

reciprocal ionic radius, which is to be expected if the ionic form of bonding is predominant. The formation constants (Table 2) are sufficiently similar in magnitude to those for glycine and alanine that these authors suggest the binding is only through the glycine-like part of the aspartate or glutamate ligand. The crystal structure of $CaG \cdot 3H_2O$ [47] supports this view, although possibly either amino acid could be tridentate with the larger ions, for reasons given in Sec. 3.1.

Formation constants have not been reported for 1:2 complexes, although such complexes can be crystallized [48]. Protonated complexes are also known [45, 48, 49], and the formation constant of $CaHG^+$ [45] corresponds better to a glutaric acid-like structure than a glycine-like one, a structure suggested by Schubert even for the deprotonated complexes [43]. The role of these and other amino acids in the creation of dental caries has been considered [50].

The formation constant of the aspartate complex of radium also has been determined [51].

3.2. Zinc and Cadmium

Surprisingly, no information is available on the mercury complexes of aspartate and glutamate, whereas reasonably detailed pictures of the chemistry of the complexes of the first two members of group 2B are available. The magnitudes of equilibrium stability constants [29, 52-60] suggest that aspartate is tridentate in the ML and ML_2^{2-} complexes, whereas glutamate is bidentate.

In addition to the simple 1:1 and 1:2 complexes, various evidence for other species has been reported. Makar and coworkers [54] have found that pH titration curves require the complexes $ZnHA^+$ and $Zn(OH)_2A_2^{4-}$. The first of these complexes has been detected by other workers [55, 29], who report $\log K_{ZnHA}^{Zn} = 1.08$ and 1.2 ± 0.2, respectively. Guevremont [55] has obtained some evidence for the formation of ZnA_3^{4-}.

Ishizuka and coworkers [29] have used ^{1}H nmr to obtain some
idea of the structure of the solution species. By assuming the
magnitudes of the vicinal coupling constants to be unaffected by
coordination of zinc, they were able to calculate the populations
of the three isomers with respect to rotation around the C_α-C_β bond
as a function of pH and zinc concentration. The results suggest
that in ZnHA$^+$ the coordination is through one carboxylate only,
whereas in ZnA and ZnA$_2^{2-}$, about 65% of the aspartate ligands are
tridentate. If the remainder are assumed to be bidentate, then
most are coordinated in a "glycine-like" manner.

The crystal structure of ZnA(H$_2$O)$_2\cdot$H$_2$O (Fig. 3) shows the
aspartate ion to be tridentate around an octahedral zinc ion [61,
62]. That of ZnG$\cdot$2H$_2$O is a distorted octahedron [63], with bonds
to the N(amino) and O(α-carboxylate) of one glutamate, to the other
α-carboxylate O of a second glutamate, to a water molecule, and to
both oxygens of the side chain carboxylate of a third glutamate.

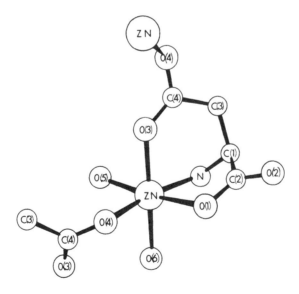

FIG. 3. Coordination sphere of the zinc atom in S-aspartatodiaquo-
zinc(II) hydrate. (Reprinted with permission from Advances in
Protein Chemistry, *22*, 257 (1967), by H. C. Freeman. Copyright by
Academic Press.)

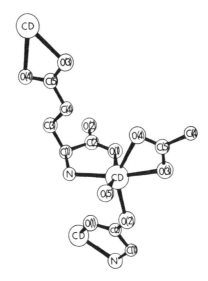

Fig. 4. Coordination sphere of the cadmium atom in S-glutamatoaquo-
cadmium(II) hydrate. (Reprinted with permission from Acta Crystallog-
raphica, *B33*, 801 (1977), by R. J. Flook, H. C. Freeman, and M. L.
Scudder. Copyright by the International Union of Crystallography.)

One of the bonds from this third glutamate is long and considerably
displaced from an axial position and the structure may alternatively
be viewed as a nearly regular square pyramid, with the zinc atom displaced 32 pm from the equatorial plane and thus almost equidistant
from the five atoms with short bonds.

    Freeman and his group have found two complexes of the formula
$CdG \cdot 2H_2O$ [64, 65]. The first, which also has been reported by Soylu
et al. [66], is isostructural with that in $ZnG \cdot 2H_2O$ and contains
cadmium in a distorted octahedral environment and glutamate in the
extended configuration [64, 65] (Fig. 4). The second, which can be
represented as $CdG(H_2O) \cdot CdG(H_2O)_2 \cdot H_2O$, again has glutamate in the
extended configuration binding to three cadmium ions. There are
two environments for cadmium, each seven ligated, with one being a
distorted pentagonal bipyramid, the other a distorted square-based
trigonal-capped coordination polyhedron [65].

3.3. Complexes with Other Main Group Metals

Soviet [67] and French [68] groups have obtained concordant stability
constant values for the glutamatogallium(III) complex. Comparison
with values for glycine or valine [67] seems to suggest that glutamate
is tridentate in this complex. The French group, using infrared spectroscopy as an aid to the interpretation of the potentiometric titrations, also identify $GaHG^{2+}$ and GaH_2G^{3+} at low pH. Crystals of the
formula $GaCl_3 \cdot AH_2$ and $GaCl_3 \cdot GH_2$ have been prepared but it seems
improbable that chelation is occuring [69].

Formation constants for the complexes $InHA^{2+}$ and $In(HA)_2^+$ at
various temperatures and ionic strengths have been determined [70],
with bidentate coordination assumed. Use of the thallous ion has
been suggested as a probe for potassium ion in biological systems,
and it is of relevance that stable complexes are formed with both
aspartic and glutamic acids [71], apparently in contrast to potassium.

With Pb^{2+}, both aspartate [72, 73] and glutamate [72, 74] form
PbL and PbL_2^{2-} complexes, although the reported values for $\log K_{ML_2}^{ML}$
are in poor agreement.

4. COMPLEXES WITH TRANSITION METALS

4.1. Cobalt(III)

The tris complexes, with the side chain carboxylates free, were
first studied by Lifschitz [75]. All four diastereomers of $Co(S-A)_3^{3-}$
were subsequently isolated as the neutral, trihydrogen species [76-
78]. There is considerable stereoselectivity, especially for the
facial isomers, which is attributed to a balance between the effects
of hydrogen bonding between the free carboxylate group and the amine
protons and steric crowding. Less marked stereoselectivity is
exhibited in the case of $Co(S-G)_3^{3-}$ [78, 79].

The bis(S-aspartato)cobaltate(III) anion has been studied
extensively and each of the three isomers characterized [80-84].

The trans(N) isomer is easily recognized by its absorption spectrum, and disagreement over assignments to the two cis(N) isomers were resolved by the determination of the crystal structure of the cis(N)-trans(O_6) isomer [85]. The bis(S-glutamato)cobaltate(III) species has not been prepared [78].

Considerable attention has been focused on mixed complexes of Co(III) with aspartate or glutamate as one component of a ternary or quaternary system. In mixed complexes in general, subtle inter- actions can lead to stereoselectivity. Because of the kinetic stability of the Co(III) complexes, it has been possible to eluci- date in detail some of the possible interactions due to functional groups in the side chains. Since mixed complexes have been con- sidered elsewhere in this series [86, 87], we discuss here primarily mixed complexes formed from aspartate or glutamate and $Co(en)_2(OH_2)_2^{3+}$ to illustrate some of these interactions with the side chain car- boxylate groups.

The reaction of S-glutamic acid with racemic $Co(en)_2(OH_2)_2^{3+}$ gives Λ- and Δ-$Co(en)_2(S-G)^+$ in equal amounts [88], although early reports claimed this not to be the case and invoked kinetic stereo- selectivity [89-92]. The perchlorate salt of the Λ diastereomer is the less soluble, and the crystal structure (Fig. 5) [91] of this salt shows the glutamate side chain to be in a conformation which supports the hypothesis of a hydrogen bond between its car- boxylate group and one of the amino protons of ethylenediamine. Whether this is the case for the Δ diastereomer or not is unknown; molecular models suggest that a linear [93] hydrogen bond can form only in the Λ diastereomer [94] of the analogous $Co(S-pn)_2A^+$ complex. It is tempting to ascribe the lower solubility of the Λ diastereomer to a lower affinity of its more compact structure for solvent mole- cules. In the case of the aspartate analogue, Kojima and Shibata [95] invoked such hydrogen bonding between an ethylenediamine NH_2 group and the side chain carboxylate of aspartate to explain the observation of two ^1H nmr signals from NH_2(ethylenediamine) in the Λ diastereomer and only one such signal in the Δ diastereomer and also the reduced rate of exchange of these protons in D_2O.

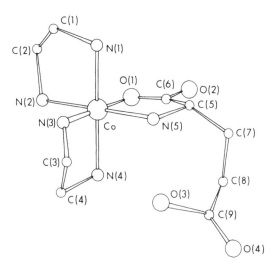

Fig. 5. The Λ-glutamatobis(ethylenediamine)cobalt(III) cation.
(Reprinted with permission from the Journal of the Chemical Society
(A), 2580 (1970), by R. D. Gillard, N. C. Payne, and G. B. Robertson.
Copyright by the Chemical Society.)

Other products from the reaction of $Co(en)_2(OH_2)_2^{3+}$ with glutamic
acid [92] have been formulated with bonding only through the two carboxylates but may be bi- or trinuclear species [96] involving bridging
glutamate ligands. Similar species with aspartate were characterized
[96].

It is well known that the methine or, in the case of glycine,
methylene protons on the α carbons of amino acids chelated to metal
ions are rather acidic and will exchange with solvent protons or
deuterons in mildly alkaline conditions. Legg and coworkers [97-101]
have studied the stereochemical course of the deuterium for protium
exchange in $Co(en)_2A^+$, $Co(en)_2G^+$, and similar compounds. The mechanism for exchange is initial attack of OD^- to form a planar,
resonance-stabilized carbanion intermediate followed by attack of
D_2O to regenerate OD^- and the amino acid. Under normal circumstances,
attack of D_2O on either side of the planar intermediate would be
expected and so racemization of the amino acid and H-D exchange
should proceed at equal rates, if the carbanion is sufficiently

long-lived. However, for both aspartic acid [98, 100] and glutamic
acid complexes [100], considerable retention of asymmetry is found.
This is ascribed to blocking of the attack of D_2O on the carbanion
by the side chain which, although now unable to form a hydrogen bond
to the ethylenediamine NH_2 protons because of the planarity at the
α carbon, has still not unfolded to a random conformation by the
time D_2O attack occurs. The retention of asymmetry was shown not
to be explainable by a difference in stability of the two diastereomers since, inter alia, the exchange in $Co(NH_3)_4A^+$ occurred with
some retention of asymmetry. In addition, exchange in the related
compounds with S-asparagine and S-glutamine gave products in which
the amino acid had completely racemized and, to confirm the validity
of the comparison, the crystal structure of Λ-Co(en)$_2$(S-asn)$^{2+}$ showed
the side chain in an extended conformation.

   Flexibility of the five-membered glycine-like ring seems to be
required for H-D exchange since the planar carbanion cannot form in
a rigid structure such as that in tridentate aspartate. In an extremely interesting paper, Legg et al. [101] showed that the tridentate aspartate in Co(dien)A^+ deuterates at the 3 carbon. The proton
exchanged was shown to be the erythro proton. Models show that the
threo proton in tridentate S-aspartate is shielded from the solution
by the two carboxylate groups whereas the erythro proton is exposed.

   Comparison of the properties of complexes of aspartate and
asparginate is often made to determine the role of the side chain
carboxylate of aspartate, since it is assumed that the amide group
in asparaginate is a poor ligand. It is interesting, therefore, to
note that the complex Co(R-A)(S-asn) has been separated as a minor
product of the reaction of $K_3Co(CO_3)_3$, S-asparagine, and R-aspartic
acid [102]. Spectral evidence suggests both ligands to be tridentate, with a facial N_3O_3 chromophore, and the complex loses a proton
between pH 2 and 9, attributed to deprotonation of the coordinated
amide group. However, the presence of a coordinated water molecule
was not entirely ruled out.

Finally, the remarkable complexes $Li_7Co(Nb_6O_{19})(S-A) \cdot 13H_2O$, $Li_7Co(Nb_6O_{14})(S-A)(OH_2) \cdot 16H_2O$, and $Li_7Co(Nb_6O_{19})(S-A)(NH_3) \cdot 13H_2O$, in which the hexaniobato ligand is tridentate, have been prepared and characterized [103].

4.2. Cobalt(II)

The most careful equilibrium stability work appears to be that of Ritsma et al. [104], who were searching for evidence of stereo-selectivity. Although the magnitude of the constants derived were such that tridentate (aspartate) and bidentate (glutamate) coordina-tion seems certain, no evidence of stereoselectivity was observed for either system. The crystal structure of $CoA \cdot 3H_2O$, which is isomorphous with $ZnA \cdot 3H_2O$ (Fig. 3), shows a tridentate aspartate ligand [62].

4.3. Nickel(II)

Crystals of $NiA \cdot 3H_2O$ are isomorphous with the zinc and cobaltous analogs [62]. That this structure, at least in gross detail, is maintained in solution is clearly shown both by the equilibrium stability constants [52, 53, 57, 59, 105-108], where values of $\log K_{NiA}^{Ni}$ are considerably greater than corresponding values for glycine, glutamate, or asparagine [104, 106-108], and by spectro-scopic investigations. Katzin and Gulyas [109] have studied the effect of solution pH and metal-ligand ratio on the electronic and circular dichroic (CD) spectra of a number of nickel-amino acid systems, and conclusions relevant to the present discussion are that aspartate coordinates as a tridentate ligand in both the 1:1 and 1:2 complexes, whereas glutamate shows behavior similar to that of alanine (among others) and so is bidentate. Surprisingly, the 1:1 asparagine complex is also thought to be tridentate, although the stability constants do not suggest this to be so. In an elegant

Study, Reilley and coworkers [110] have measured the ^{1}H nmr contact shifts of protons in octahedral nickel-amino acid complexes. The striking difference in the spectra of the aspartic acid and glutamic acid complexes is the very large downfield shift (-145 ppm) of the methine proton of aspartate compared to that in glutamate (-45 ppm) which requires an approximately equatorial position for the methine proton in the aspartate complex [62]. The shifts for the various methylene groups are also consistent with tridentate aspartate and bidentate glutamate, with the methylene resonance of aspartate upfield from those of glutamate. In the mixed complexes $Ni(EDDA)A^{2-}$ and $Ni(EDDA)G^{2-}$, where only two coordination positions are available for the amino acid ligand, the spectrum of each is similar to that of glutamate in $NiG(H_2O)_4$.

Two groups [104, 111] have measured the formation constant of $Ni(S-A)(R-A)^{2-}$; as with the analogous cobalt(II) complex and the bis(glutamato)nickel(II) complex [106], there is no evidence of stereoselectivity. The mixed complexes of NiA and NiG with glycinate show a small stabilization relative to the parent complexes, NiA_2^{2-} and $Ni(gly)_2$, probably because of the charge difference [59, 107, 108]. In addition, the complexes $NiA(gly)_2^{2-}$, $NiG_2(gly)^{3-}$, and $NiG(gly)_2^{2-}$ were detected in greater than statistical concentration but the complex $NiA_2(gly)^{3-}$ was not found [59]. The protonated complexes $NiHA^+$ and $CuHA^+$ have been reported recently by Gergely and coworkers [59, 107, 108].

The kinetics of formation of NiA have been studied by stopped-flow methods [112], although this work has been criticized recently on the grounds that the zwitterion was wrongly considered to be unreactive [113, 114]. The six-membered ring in NiA has been shown to be more labile than the five-membered ring by comparison of the rates of substitution of NiA and Ni(NTA) by the ligands diethylene-triaminepentaacetic acid and cyclohexane-1,2-diamine-N,N,N',N'-tetra-acetic acid [115].

4.4. Copper (II)

As usual, evidence suggests that aspartate is tridentate and glutamate bidentate. Thus, the crystal structure of $CuG \cdot 2H_2O$ [116] is
similar to that of one of the structures observed for $CdG \cdot 2H_2O$
(Fig. 4) [64-66], where the side chain is extended. The formation
constant of CuA is considerably greater than those for Cu-gly,
Cu-norvalinate, or CuG [27], and the CD spectra of CuA_2^{2-} [117, 118]
and of $CuA(gly)^-$ [119] provide evidence for axial binding (presumably
of the side chain carboxylate); this CD evidence is less compelling
for the glutamate analogs. In CuA_2^{2-}, it may be that only one of the
side chain carboxylates is bound since the cumulative stability constant of $CuAG^{2-}$ is close to that of CuA_2^{2-} [27]; similar conclusions
were reached from the CD measurements [117, 118]. Kirschner [120]
prepared $K_2CuA_2 \cdot H_2O$, and the ir spectrum of the crystals shows only
a single absorption in the carbonyl region at a value (1,620 cm^{-1})
consistent with metal-coordinated carboxylates: The conclusion was
that both aspartates are tridentate in the crystal, although we
wonder about the role of the water molecule.

Copper complexes of the protonated ligands have been of considerable interest. In very early work, Albert [52] experienced
difficulty in calculating values of $K_{CuL_2}^{Cu}$ at low pH, and several
recent reports have shown the presence of one or more protonated
complexes in addition to CuL and CuL_2^{2-} [27, 107, 121, 122]. Brookes
and Pettit [122] claimed to identify $CuHL^+$, CuH_2L_2, and $CuHL_2^-$,
whereas earlier potentiometric work [27, 121] found only the first
of these. Kirschner [120] isolated $CuH_2A_2 \cdot 1/2H_2O$, which had absorp-
tions in the ir typical of $-CO_2H$ groups. The $pK_{CuHA^+}^H$ is close to
$pK_{H_2A}^H$ of the free ligand [27, 121], but since these species have
different charges it is difficult to compare them directly. For
the glutamate system, comparisons can be made that support Ritsma's
contention that structure III is more likely than structure IV for
$CuHG^+$ [121]. First, comparison of the ionization constant of $CuHG^+$

III   IV

with the microscopic k_2 shows that $CuHG^+$ is a weaker acid, as is
reasonable. Second, the microscopic formation constants calculated
for III and IV from K^{Cu}_{CuHG} and the microconstants in Table 1 by
methods described previously [123] are more in line with III. The
microscopic formation constant of III, $k_{III} = [CuHG^+]/[Cu^{2+}][f]$
where tautomer f is identified in Fig. 2, is equal to $K^{Cu}_{CuHG^+}/\alpha_f$
where α_f is that fraction of HG^- that is in the form of tautomer f.
From the microconstants in Table 1, α_f is calculated to be $1.35 \times$
10^{-5}, which gives $k_{III} = 10^{7.85}$ using $K^{Cu}_{CuHG} = 10^{2.98}$ [122]. The
microscopic constant of IV, $k_{IV} = [CuHG^+]/[Cu^{2+}][e]$ where e is also
identified in Fig. 2, can be taken to be essentially equal to K^{Cu}_{CuHG}
because e is the predominant tautomer of HG^-. Since k_{III} and k_{IV}
are in terms of the specific forms of the ligand in the proposed
complexes, they can be compared directly with the formation constants of model complexes to aid in the identification of the bonding
scheme in $CuHG^+$. For this purpose, the formation constants for the
glycinate and norvalinate complexes of Cu^{2+} are both $10^{8.07}$ [27],
and that for the Cu^{2+} complex with the dianion of glutaric acid,
$HO_2C(CH_2)_3CO_2H$, is $\sim 10^{2.4}$ [124] at conditions similar to those in
the work being discussed. If $CuHG^+$ has structure IV, k_{IV} would be
expected to be less than the formation constant for the glutaric
acid complex because of the positive charge on the ammonium group
in $CuHG^+$, whereas k_{III} agrees quite well with the formation constants for glycinate and norvalinate.

    Contrary to earlier reports [125], there appears to be no
appreciable stabilization of $Cu(S-A)(R-A)^{2-}$ over $Cu(S-A)_2^{2-}$ [121] or

$Cu(S-G)(R-G)^{2-}$ over $Cu(S-G)_2^{2-}$ [106]. Assuming the structure of CuA_2^{2-} to involve apical coordination, it seems that there is no marked preference for cis or trans coordination in the equatorial plane, as might be expected, for reasons discussed in Sec. 1. Gergely and coworkers [27] found appreciable stabilization of the ternary $CuA(gly)^-$ complex, none for the $CuG(gly)^-$ complex, and a small stabilization of the $CuAG^{2-}$ complex relative to the parent complexes. These results are not easy to explain. However, the variation in stability constant with ionic strength supports the assumption that the $CuA(gly)^-$ complex is stabilized because of the charge difference relative to the parent complexes [27, 107]. Simple dipeptides are tridentate to Cu^{2+} at sufficiently high pH, with binding through the N(amino), N(deprotonated amide), and carboxylate group. The mixed complexes of glygly, gly-R,S-ala, and R,S-ala-R,S-ala with a number of amino acids have been reported [126] and the high stability of the ternary complex $Cu(dipep(-H))A^{2-}$ has been interpreted as showing coordination of the side chain carboxylate of aspartate in an axial position.

Most interesting are the ternary complexes where some electrostatic interaction, stabilizing or destabilizing, can occur between side chain groups. As discussed in the Introduction, geometrical isomerism may have an important effect on these systems. Brookes and Pettit [127] have measured the formation constants of Cu(S- or R-his) $(S-A)H_x$ and Cu(S - or R-his)$(S-G)H_x$, where x = 0, 1, 2. They assume cis coordination of the two amino groups and apical coordination of the side chain carboxylate of A, and show that the formation constant of Cu(R-his)(S-A) is greater than that of Cu(S-his)(S-A), where the two carboxylate groups are on the same side of the tetragonal plane. In the case of glutamate, however, where there is no apical binding of the side chain carboxylate, no stereoselectivity is found. These observations have been extended [122]. The converse observation, stereoselectivity through an attractive mechanism, has also been reported [128, 129]. The mixed complexes between Cu^{2+}, each enantiomer of aspartic or glutamic acid, and S-arginine, S-lysine, and

S-ornithine have been isolated, and the far-ir spectra suggest trans coordination in the S,S pairs and cis coordination in the R,S pairs, thus placing the two side chains on the same side of the plane [130]. This observation has been used recently [129] to achieve partial resolution of R,S-aspartic or R,S-glutamic acids by crystallizing from solution the mixed compound with the optically pure forms of the basic ligands listed above. The complex containing the R enantiomer of the aminodicarboxylic acid crystallized preferentially.

The rate for ligand exchange in a number of bis(amino-acidato)copper(II) complexes has been determined by ^{1}H nmr by measurement of the broadening of the water signal [130]. The rate for glutamate is lower than for most amino acids, probably because of the double-negative charge on both the complexes and the reacting ligand. That for aspartate is still lower because of the rate-decreasing effect of the axial coordination in one of the two bound aspartate ligands.

4.5. Other First Row Transition Metals

With chromium, both CrA_2^- and CrA_3^{3-} complexes have been isolated; the latter presumably has free side chain carboxylate groups. All three isomers of $Cr(S-A)_2^-$ have been separated, but configurational assignments have not been made [131]. The complex $NH_4[Cr(S,R-A)_2]$ has been isolated [132] and stated to be a mixture of $Cr(S-A)_2^-$ and $Cr(R-A)_2^-$ in the cis(N)-trans(O_5) geometrical form (Fig. 1), but there is a lack of agreement as to the absorption spectrum with that of Weiss [131]. The tris complex has also been prepared [133-135]. It appears that the stoichiometry can be controlled by pH, with the tris complex forming at pH < 2 [135-137]. Some attempts have been made to measure formation constants of Cr(III)-aspartic acid complexes but it is doubtful if equilibrium conditions were reached [136, 137]. The glutamic acid complexes of chromium have not been well characterized [134]. It is interesting that the glucose tolerance factor contains chromium, and a synthetic material with similar properties has been made by refluxing chromium acetate with nicotinic acid, glycine, cysteine, and glutamic acid [138].

Ferric ion [139, 140] forms strong complexes that show the unusual feature that log K^{Fe}_{FeG} is greater than log K^{Fe}_{FeA}; in each case the formation of a tridentate complex is invoked, with support from the markedly lower values obtained with all other amino acids including asparagine. Complexes also form with ferrous ion [141], and here the magnitude of the formation constants indicate that the ligand is probably tridentate in the aspartate complex and bidentate in the glutamate complex.

With manganese, the complex $MnA(H_2O)_2$ has been isolated [142], and a value for K^{Mn}_{MnA} has been reported [52, 143]. Similar studies with glutamic acid have been reported [52]. Aspartic acid forms rather stable 1:1 and 1:2 complexes with vanadyl ion [144].

4.6. Second and Third Row Transition Metals

Little work has been reported on the aspartate and glutamate complexes of these metal ions, whose aqueous solution chemistry is often dominated by their easy hydrolysis and/or kinetic inertness.

The extensive hydrolysis of Mo(VI) makes study of its complexes rather difficult [145]. At low pH, isopolymolybdates form while at high pH the Mo(VI) is present as the stable MoO_4^{2-} ion; around pH 6 there is a narrow "window" where complexation can be studied. Because of the unsymmetrical nature of the equations for formation of isopoly-compounds, the extent of complex formation depends more strongly than usual on metal ion concentration. However, chiroptical spectra [39, 146], pH titrations [39], and preparative work [147] show the presence of an aspartate complex of the MoO_3 core, MoO_3A^{2-}. The cumulative formation constant for the reaction

$$MoO_4^{2-} + 2H^+ + A^{2-} \rightleftarrows MoO_3A^{2-} + H_2O$$

is 6.23×10^{16}, and that for glutamic acid is 6.11×10^{16} [39]. These results suggest that both ligands are tridentate.

Differing opinions [148, 149] on the question of the formation of similar complexes with W(VI) have been resolved by Rabenstein et al. [39], who were unable to detect any complex formation with either amino acid. The more extensive formation of isopolycompounds of W(VI), relative to Mo(VI), leaves virtually no window in which complexes can form.

Pd(II) appears to form 1:1 and 1:2 chelates with aspartate and glutamate and concordant values for the stability constants for the aspartate complexes have been reported by two groups [150, 151]. The side chains appear not to be coordinated [152] as expected for square planar coordination and, in the case of the glutamate complex, have been titrated to the phenolphthalein end point [153].

With Pt(II), both chelated and monodentate aspartic acid complexes have been prepared; ring closure in $PtCl_2A_2^{4-}$ is quite slow, but both cis and trans isomers of PtA_2^{2-} should be accessible [154]. For the compound of trans geometry, and its glutamate analog, the magnitude of the $^{195}Pt-^1H$ (methine) coupling constant shows that the chelate ring has a conformation in which the side chain and its carboxylate group are directed away from the metal atom. The methine proton exchanges with deuterium at alkaline pD at a rate comparable to the rate of racemization of the amino acid [155, 156], consistent with the extended nature of the side chain and the bidentate chelation.

With rhodium, both $H_3RhA_3 \cdot 2H_2O$ and $H_3RhG_3 \cdot 2H_2O$ have been prepared by refluxing $RhCl_3 \cdot 3H_2O$ and the appropriate amino acid [157]. The ligands are presumably bound through the five-membered ring. Tridentate glutamate has been proposed in $RhCl(H_2O)_2G$, although this compound may be polymeric [158].

5. COMPLEXES WITH LANTHANIDES AND ACTINIDES

5.1. The Lanthanides

The coordination chemistry of these elements as a group is similar to that of calcium, strontium, and barium, and dependent more upon

electrostatic interactions and ionic size than upon electronic interactions. Thus, the strongest binding sites on the aspartate and
glutamate ligands are the carboxylate groups. The amino groups,
even in the deprotonated form, are not expected to interact strongly
with the metal ion [159].

The complexation of lanthanide ions and amino acids has been
studied primarily by uv-vis and CD spectroscopy, and by pH titration
methods. Unfortunately, most of the various studies in which formation constants have been determined disagree on the nature of the
species and appear to neglect the problem of metal ion hydrolysis,
which occurs at quite low pH. Martin and coworkers [160] have shown
that the pH titration of solutions containing a mixture of one of many
amino acids and a Ln(III) ion is characterized by titration as though
the separate components titrated alone. At low pH the carboxyl proton
is titrated, near pH 7 to 9 the metal ion undergoes hydrolysis consuming 2.4 to 2.8 equivalents of base, and finally the amino group is
deprotonated exactly as if the ligand was in solution alone. The
hydrolysis process absorbs about 2.5 equivalents of OH^-, independent
of the quantity of amino acid, presumably forming a polynuclear
bridged (or polymeric) species such as $M_2(OH)_5L_2$. The pH titration
of aspartic acid with Ln(III) is essentially identical to that of
Ln(III) alone except that only two equivalents of OH^- are consumed
during hydrolysis. This suggests that the aspartic acid ligand is
bidentate through both carboxyl groups and forms a sufficiently
stable and soluble complex that normal hydrolysis is retarded. A
species [perhaps mononuclear, e.g., $M(OH)_2A$] may form in solution.
Glutamic acid, however, forms a precipitate and presumably behaves
in much the same way as do more simple amino acids, perhaps binding
at the α or side chain carboxylate.

A study of the reaction of Pr(III) with amino acids revealed
three types of behavior in the CD spectrum based on ratios of ligand
to metal and on the pH. The spectra of the aspartic acid and glutamic
acid complexes were interpreted as reflecting chelation through two
carboxylate groups [161].

5.2. The Actinides

There has been very little reported about the coordination of
aspartic and glutamic acids to the actinides. The precipitates
from solutions of thorium and uranium with aspartic and glutamic
acid have been studied [162]. In the precipitates of uranium complexes, the ammonium group of the amino acids remained protonated,
suggesting that coordination was exclusively through the carboxylate
groups. Feldman and Koval [163] came to the conclusion from spectro-
copic and potentiometric data that UO_2^{2+} binds strongly to the carboxylate group but only weakly to the amine group. The complexes of
aspartic and glutamic acids were approximately equal in stability
to those of acetate and succinate suggesting the absence of chelation. It was also suggested that the carboxylate group occupies
two coordination sites of UO_2^{2+} because the UO_2^{2+} complex of acetate
is about twice as strong as the acetate complex of the 10 rare
earths studied [164].

ACKNOWLEDGMENTS

We wish to thank the publishers and authors of Refs. 62, 64, 91, and
132 for permission to reproduce figures from their publications. It
is also a pleasure to thank our typist, Miss A. Wiseman, draftsper-
sons J. Stellick and F. Boychuk, and librarians Ms. B. Boyd and
Mrs. T. Williams for their invaluable assistance.

ABBREVIATIONS AND DEFINITIONS

A	aspartate, $C_4H_5O_4N$	G	glutamate, $C_5H_7O_4N$
L	either of A or G	M	a metal ion
asn	asparaginate, $C_4H_7O_3N_2$	gly	glycinate, glycyl
ala	alaninate, alanyl	his	histidinate
en	1,2-diaminoethane	pn	1,2-diaminopropane

nmr	nuclear magnetic resonance	ppm	parts per million
pot.	potentiometric titration (glass electrode)	Cd	potentiometric titration (cadmium electrode)
pol.	polarography	CD	circular dichroism
ix	ion exchange	sol.	solubility

$$K_{H_nL}^H = \frac{aH^+ \cdot [H_{n-1}L]}{[H_nL]} \qquad K_{ML}^M = \frac{[ML]}{[M][L]}$$

$$K_{ML_2}^{ML} = \frac{[ML_2]}{[ML][L]} \qquad \beta pqr = \frac{[M_pH_qL_r]}{[M]^p[H]^q[L]^r}$$

REFERENCES[*]

1. H. B. Vickery and C. L. A. Schmidt, *Chem. Rev., 9,* 169 (1931).

2. L. Pasteur, *Ann. Chim. Phys., 34,* 30 (1852).

3. H. Kolbe, *Ann., 121,* 232 (1862).

4. H. Ritthausen, *J. Prakt.Chem., 103,* 233 (1868).

5. Ibid., *107,* 218 (1869).

6. L. F. Burroughs, The Biochemistry of Fruits and their Products (A. C. Hulme, ed.), Academic Press, New York, 1970, p. 119ff.

7. J. C. Dickinson, H. Rosenblum, and P. B. Hamilton, *Pediatrics, 36,* 2 (1965).

8. H. Ritthausen, *J. Prakt. Chem., 99,* 454 (1869).

9. Ibid., *106,* 445 (1869).

10. Ibid., *103,* 239 (1868).

11. Ibid., *107,* 218 (1866).

12. Absolute Configuration of Metal Complexes, Wiley-Interscience, New York, 1971, p. 1ff.

13. C. J. Hawkins, ibid., p. 132ff.

14. IUPAC, *Inorg. Chem., 9,* 1 (1970).

15. J. T. Edsall and J. Wyman, Biophysical Chemistry, 2nd ed., Academic Press, New York, 1958, p. 494.

16. A. Neuberger, *Biochem. J., 30,* 2085 (1936).

17. T. L. Hill, *J. Phys. Chem., 48,* 101 (1944).

---

[*]Russian language journals were consulted in their English translations. The page number in the original is given in square brackets.

18. T. L. Hill, *J. Chem. Phys., 12,* 56 (1944).

19. J. T. Edsall and M. H. Blanchard, *J. Amer. Chem. Soc., 55,* 2337 (1933).

20. R. E. Benesch and R. Benesch, *J. Amer. Chem. Soc., 77,* 5877 (1955).

21. D. L. Rabenstein and T. L. Sayer, *Anal. Chem., 48,* 1141 (1976).

22. T. L. Sayer and D. L. Rabenstein, *Can. J. Chem., 54,* 3392 (1976).

23. J. Llopis and D. Ordonez, *J. Electroanal. Chem., 5,* 129 (1963).

24. E. R. B. Smith and P. K. Smith, *J. Biol. Chem., 146,* 187 (1942).

25. A. C. Batchelder and C. L. A. Schmidt, *J. Biol. Chem., 90,* 165 (1931).

26. R. D. Graham, D. R. Williams, and P. A. Yeo, *J. Chem. Soc. Perkin II,* 1876 (1972).

27. I. Nagypál, A. Gergely, and E. Farkas, *J. Inorg. Nucl. Chem., 36,* 699 (1974).

28. J. J. Christensen, R. M. Izatt, and L. D. Hansen, *J. Amer. Chem. Soc., 89,* 213 (1967).

29. H. Ishizuka, T. Yamamoto, Y. Arata, and S. Fujiwara, *Bull. Chem. Soc. Jap., 46,* 468 (1973).

30. F. Taddei and L. Pratt, *J. Chem. Soc.,* 1553 (1964).

31. K. G. R. Pachler, *Spectrochim. Acta, 20,* 581 (1964).

32. K. G. R. Pachler, *Z. Anal. Chem., 244,* 211 (1967).

33. M. Kainosho and K. Ajisaka, *J. Amer. Chem. Soc., 97,* 5630 (1975).

34. S. Fujiwara, H. Ishizuka, and S. Fudano, *Chem. Lett.,* 1281 (1974).

35. J. L. Derissen, H. J. Endeman, and A. F. Peerdeman, *Acta Crystallogr., B24,* 1349 (1968).

36. S. T. Rao, *Acta Crystallogr., B29,* 1718 (1973).

37. S. Hirokawa, *Acta Crystallogr., 8,* 637 (1955).

38. F. L. Siegel, Calcium-Binding Proteins, *Structure and Bonding, 17,* 221 (1973).

39. D. L. Rabenstein, M. S. Greenberg, and R. Saetre, *Inorg. Chem., 16,* 1241 (1977).

40. T. L. James and J. H. Noggle, *Bioinorganic Chem., 2,* 69 (1972).

41. R. F. Lumb and A. E. Martell, *J. Phys. Chem., 57,* 690 (1953).

42. I. Greenwald, *J. Phys. Chem., 43,* 379 (1939).

43. J. Schubert, *J. Amer. Chem. Soc., 76,* 3442 (1954).

44. J. Schubert and A. Lindenbaum, *J. Amer. Chem. Soc.*, *74*, 3529 (1952).

45. C. W. Davies and G. M. Waind, *J. Chem. Soc.*, 301 (1950).

46. M. Hardel, *Hoppe-Seyler's Z. Physiol. Chem.*, *346*, 224 (1966).

47. H. Einspahr and C. E. Bugg, *Acta Crystallogr.*, *B30*, 1037 (1974).

48. W. K. Anslow and H. King, *Biochem. J.*, *21*, 1168 (1927).

49. N. Masayuki and M. Kobayashi, *Yakugaku Zasshi*, *88*, 1128 (1968).

50. T. Morch, I. Punwani, and E. Greve, *Caries Res.*, *5*, 135 (1971).

51. J. Schubert, E. R. Russell, and L. S. Myers, *J. Biol. Chem.*, *185*, 387 (1950).

52. A. Albert, *Biochem. J.*, *50*, 690 (1952).

53. D. J. Perkins, *Biochem. J.*, *55*, 649 (1953).

54. G. K. R. Makar, M. L. D. Touche, and D. R. Williams, *J. Chem. Soc. Dalton*, 1016 (1976).

55. R. Guevremont, Ph.D. Thesis, University of Alberta, 1978.

56. S. Chaberek and A. E. Martell, *J. Amer. Chem. Soc.*, *74*, 6021 (1952).

57. R. Munze, A. Guthert, and H. Matthes, *Z. Phys. Chem.*, Leipzig, *241*, 240 (1969).

58. G. J. M. Heijne and W. E. van der Linden, *Talanta*, *22*, 923 (1975).

59. A. Gergely and E. Farkas, *Magy. Kem. Foly.*, *81*, 471 (1975).

60. G. Reinhard, R. Dreyer, and R. Munze, *Z. Phys. Chem.*, Leipzig, *254*, 226 (1973).

61. T. Doyne, R. Pepinsky, and T. Watanabe, *Acta Crystallogr.*, *10*, 438 (1957).

62. T. Doyne, Ph.D. thesis, Pennsylvania State University, 1957, as cited in H. C. Freeman, *Adv. Protein Chem.*, *22*, 257 (1967).

63. C. M. Gramaccioli, *Acta Crystallogr.*, *21*, 600 (1966).

64. R. J. Flook, H. C. Freeman, and M. L. Scudder, *Acta Crystallogr.*, *B33*, 801 (1977).

65. R. J. Flook, H. C. Freeman, C. J. Moore, and M. L. Scudder, *J. Chem. Soc., Chem. Comm.*, 753 (1973).

66. H. Soylu, D. Ulku, and J. C. Morrow, *Z. Kristallogr.*, *140*, 281 (1974).

67. E. A. Zekharova and V. N. Kumok, *J. Gen. Chem. USSR*, *38*, 1868 [1922] (1968).

68. P. Bianco, J. Haladjian, and R. Pilard, *J. Chim. Phys.*, *73*, 280 (1976).

69. I. A. Sheka and K. I. Arsenin, *J. Gen. Chem. USSR, 44,* 2286 [2332] (1974).

70. R. Sarin and K. N. Munshi, *J. Inorg. Nucl. Chem., 34,* 581 (1972).

71. F. Ya. Kul'ba, V. G. Ushakova, and Yu. B. Yakovlev, *Russ. J. Inorg. Chem., 20,* 43 [79] (1975).

72. G. N. Rao and R. S. Subrahamanya, *Proc. Indian Acad. Sci., 60A,* 185 (1964).

73. M. Kodama and S. Takahashi, *Bull. Chem. Soc. Jap., 44,* 697 (1971).

74. M. Kodama, *Bull. Chem. Soc. Jap., 47,* 1547 (1974).

75. I. Lifschitz and W. Froentjes, *Rec. Trav. Chim., 60,* 225 (1941).

76. M. Shibata, H. Nishikawa, and K. Hosaka, *Bull. Chem. Soc. Jap., 40,* 236 (1967).

77. K. Kawasaki, J. Yoshii, and M. Shibata, ibid., *43,* 3819 (1970).

78. K. Kawasaki and M. Shibata, ibid., *45,* 3100 (1972).

79. R. D. Gillard and N. C. Payne, *J. Chem. Soc.(A),* 1197 (1969).

80. J. I. Legg and D. W. Cooke, *J. Amer. Chem. Soc., 89,* 6854 (1967).

81. K. Hosaka, H. Nishikawa, and M. Shibata, *Bull. Chem. Soc. Jap., 42,* 277 (1969).

82. L. R. Froebe, S. Yamada, J. Hidaka, and B. E. Douglas, *J. Coord. Chem., 1,* 183 (1971).

83. S. Yamada, J. Hidaka, and B. E. Douglas, *Inorg. Chem., 10,* 2187 (1971).

84. J. I. Legg and J. A. Neal, ibid., *12,* 1805 (1973).

85. I. Oonishi, S. Sato, and Y. Saito, *Acta Crystallogr., B31,* 1318 (1975).

86. H. Sigel (ed.), Metal Ions in Biological Systems, Vol. 2, Marcel Dekker, New York, 1973.

87. This volume, Chap. 6 by L. D. Pettit and R. J. W. Hefford

88. D. A. Buckingham, J. Dekker, A. M. Sargeson, and L. G. Marzilli, *Inorg. Chem., 12,* 1207 (1973).

89. J. H. Dunlop, R. D. Gillard, and N. C. Payne, *J. Chem. Soc. (A),* 1469 (1967).

90. J. H. Dunlop, R. D. Gillard, N. C. Payne, and G. B. Robertson, *Chem. Comm.,* 874 (1966).

91. R. D. Gillard, N. C. Payne, and G. B. Robertson, *J. Chem. Soc. (A),* 2579 (1970).

92. R. D. Gillard, R. Maskill, and A. Pasini, *J. Chem. Soc. (A),* 2268 (1971).

93. I. Olovsson and P.-G. Jonsson, in The Hydrogen Bond (P. Schuster, G. Zundel, and C. Sandorfy, eds.), North Holland, Amsterdam, p. 395ff.

94. Y. Kojima and M. Shibata, *Inorg. Chem.*, *10*, 2382 (1971).

95. Ibid., *12*, 1009 (1973).

96. T. Yasui, H. Kawaguchi, Z. Kanda, and T. Ama, *Bull. Chem. Soc. Jap.*, *47*, 2393 (1974).

97. J. I. Legg and J. Steele, *Inorg. Chem.*, *10*, 2177 (1971).

98. W. E. Keyes and J. I. Legg, *J. Amer. Chem. Soc.*, *95*, 3431 (1973).

99. Ibid., *98*, 4970 (1976).

100. W. E. Keyes, R. E. Caputo, R. D. Willett, and J. I. Legg, ibid., *98*, 6939 (1976).

101. J. A. McLarin, L. A. Dressel, and J. I. Legg, ibid., *98*, 4150 (1976).

102. H. Takenaka and M. Shibata, *Chem. Lett.*, 535 (1976).

103. Y. Hosokawa, J. Hidaka, and Y. Shimura, *Bull. Chem. Soc. Jap.*, *48*, 3175 (1975).

104. J. H. Ritsma, G. A. Wiegers, and F. Jellinek, *Rec. Trav. Chim.*, *84*, 1577 (1965).

105. D. I. Ismailov, A. P. Borisova, I. A. Saoich, and V. I. Spitsyn, *Dokl. Akad. Nauk SSSR*, *207*, 976 [651] (1972).

106. D. S. Barnes and L. D. Pettit, *J. Inorg. Nucl. Chem.*, *33*, 2177 (1971).

107. A. Gergely, I. Nagypál, and E. Farkas, *Magy. Kem. Foly.*, *80*, 56 (1974).

108. A. Gergely, I. Nagypál, and E. Farkas, *Acta Chim. Acad. Sci. Hung.*, *82*, 43 (1974).

109. L. I. Katzin and E. Gulyas, *J. Amer. Chem. Soc.*, *91*, 6940 (1969).

110. F. F.-L. Ho, L. E. Erickson, S. R. Watkins, and C. N. Reilley, *Inorg. Chem.*, *9*, 1139 (1970).

111. J. R. Blackburn and M. M. Jones, *J. Inorg. Nucl. Chem.*, *35*, 1605 (1973).

112. J. C. Cassatt and R. G. Wilkins, *J. Amer. Chem. Soc.*, *90*, 6045 (1968).

113. J. E. Letter and R. B. Jordan, ibid., *97*, 2381 (1975).

114. R. H. Voss and R. B. Jordan, ibid., *98*, 2173 (1976).

115. M. Kodama and T. Ueda, *Bull. Chem. Soc. Jap.*, *43*, 419 (1970).

116. C. M. Gramaccioli and R. E. Marsh, *Acta Crystallogr.*, *21*, 594 (1966).

117. J. M. Tsangaris and R. B. Martin, *J. Amer. Chem. Soc., 92,* 4255 (1970).

118. E. W. Wilson, M. H. Kasperian, and R. B. Martin, *J. Amer. Chem. Soc., 92,* 5365 (1970).

119. K. M. Wellman, T. G. Mecca, W. Mungall, and C. R. Hare, *J. Amer. Chem. Soc., 90,* 805 (1968).

120. S. Kirschner, *J. Amer. Chem. Soc., 78,* 2372 (1956).

121. J. H. Ritsma, *Rec. Trav. Chim., 94,* 210 (1975).

122. G. Brookes and L. D. Pettit, *J. Chem. Soc. Dalton,* 1918 (1977).

123. D. L. Rabenstein, M. S. Greenberg, and C. A. Evans, *Biochemistry, 16,* 977 (1977).

124. L. G. Sillén and A. E. Martell, Chemical Society Special Publication No. 17, 1964, p. 284.

125. W. E. Bennett, *J. Amer. Chem. Soc., 81,* 246 (1959).

126. I. Nagypál and A. Gergely, *Magy. Kem. Foly., 82,* 448 (1976).

127. G. Brookes and L. D. Pettit, *J. Chem. Soc., Chem. Comm.,* 385 (1975).

128. T. Sakurai, O. Yamauchi, and A. Nakahara, *Bull. Chem. Soc. Jap., 49,* 169 (1976).

129. T. Sakurai, O. Yamauchi, and A. Nakahara, *J. Chem. Soc., Chem. Comm.,* 553 (1976).

130. I. Nagypál, E. Farkas, and A. Gergely, *J. Inorg. Nucl. Chem., 37,* 2145 (1975).

131. G. Grouchi-Witte and E. Weiss, *Z. Naturforsch., 31B,* 1190 (1976).

132. H. Oki and Y. Takahashi, *Bull. Chem. Soc. Jap., 50,* 2228 (1977).

133. L. M. Volshtein, G. G. Motyagina, and L. S. Anokhova, *J. Inorg. Chem., USSR, 1,* 215 [2378] (1956).

134. H. Mizuochi, A. Uehera, E. Kyuno, and R. Tsuchiya, *Bull. Chem. Soc. Jap., 44,* 1555 (1971).

135. A. Lassocinska, *Rocz. Chem., 47,* 889 (1973).

136. A. Lassocinska, *Rocz. Chem., 48,* 867 (1974).

137. H. Mizuochi, S. Shirakata, E. Kyuno, and R. Tsuchiya, *Bull. Chem. Soc. Jap., 43,* 397 (1970).

138. E. W. Toepfer, W. Mertz, M. M. Polansky, E. E. Roginski, and W. R. Wolf, *J. Agr. Food Chem., 25,* 162 (1977).

139. C. V. Smythe and C. L. A. Schmidt, *J. Biol. Chem., 88,* 241 (1930).

140. D. D. Perrin, *J. Chem. Soc.,* 3125 (1958).

141. D. D. Perrin, *J. Chem. Soc.*, 290 (1959).

142. L. P. Berezina, A. I. Pozigun, and V. L. Misyurenko, *Russ. J. Inorg. Chem.*, *15*, 1244 [2402] (1970).

143. H. Kroll, *J. Amer. Chem. Soc.*, *74*, 2034 (1952).

144. S. G. Tak, O. P. Sunar, and C. P. Trivedi, *J. Indian Chem. Soc.*, *49*, 121 (1972).

145. H. Eguchi, T. Takeuchi, A. Ouchi, and A. Furuhashi, *Bull. Chem. Soc. Jap.*, *42*, 3585 (1969).

146. M. D. Azevedo, R. G. Costa, and M. T. Vilhena, *Rev. Port. Quim.*, *15*, 35 (1973).

147. R. J. Butcher, H. K. J. Powell, C. J. Wilkins, and S. H. Yong, *J. Chem. Soc., Dalton Trans.*, 356 (1976).

148. M. K. Singh and M. N. Srivastava, *J. Inorg. Nucl. Chem.*, *34*, 2081 (1972).

149. D. H. Brown, and D. Neumann, *J. Inorg. Nucl. Chem.*, *37*, 330 (1975).

150. O. P. Sunar and C. P. Trivedi, ibid., *33*, 3990 (1971).

151. M. K. Singh and M. N. Srivastava, *J. Inorg. Nucl. Chem.*, *34*, 2067 (1972).

152. E. W. Wilson and R. B. Martin, *Inorg. Chem.*, *9*, 528 (1970).

153. P. Spacu and I. Scherzer, *Z. Anorg. Allg. Chem.*, *319*, 101 (1962).

154. L. M. Volshtein and L. S. Anokhova, *Russ. J. Inorg. Chem.*, *6*, 152 [300] (1961).

155. L. E. Erickson, J. W. McDonald, J. K. Howie, and R. P. Clow, *J. Amer. Chem. Soc.*, *90*, 6371 (1968).

156. L. E. Erickson, A. J. Dappen, and J. C. Uhlenhopp, *J. Amer. Chem. Soc.*, *91*, 2510 (1969).

157. H. Frye, C. Luschak, and D. Chinn, *Z. Naturforsch.*, *22B*, 268 (1967).

158. H. Kalberer and H. Frye, *Inorg. Nucl. Chem. Lett.*, *7*, 215 (1971).

159. T. Moeller, D. F. Martin, L. C. Thompson, R. Ferrus, G. R. Feistel, and W. J. Randall, *Chem. Rev.*, *65*, 1 (1965).

160. R. Prados, L. G. Stadtherr, H. Donato, and R. B. Martin, *J. Inorg. Nucl. Chem.*, *36*, 689 (1974).

161. L. I. Katzin, *Coord. Chem. Rev.*, *5*, 279 (1970).

162. G. M. Sergeev and J. A. Korshunov, *Soviet Radiochemistry*, *16*, 771 [787] (1974).

163. I. Feldman and L. Koval, *Inorg. Chem.*, *2*, 145 (1963).

164. A. Sonesson, *Acta, Chem. Scand.*, *12*, 1937 (1958).

Chapter 3

THE COORDINATION CHEMISTRY OF L-CYSTEINE
AND D-PENICILLAMINE

A. Gergely and I. Sóvágó
Department of Inorganic and Analytical Chemistry
Kossuth Lajos University
Debrecen, Hungary

1. INTRODUCTION

The properties and more important chemical reactions of the naturally
occurring amino acid L-cysteine and its derivative, D-penicillamine,
have already been subjected to wide-reaching investigation. The most
important reactions are the processes involving participation of the

mercapto group; the main biochemical aspects of these have been
reviewed by Jocelyn [1]. The complex chemical significance of both
cysteine and penicillamine is determined by the mercaptosulfur donor
atom, which is very soft in character. Via the mercapto group, these
compounds may participate in both redox and acid-base-type reactions.
Extremely great biological importance is attached to the study of the
metal-sulfur bonds formed in such processes, primarily in the nonheme
iron proteins [2] and the blue copper proteins [3].

The significance of D-penicillamine is enhanced by the fact
that it displays independent therapeutic activity [4]. Its most
valuable use is for the treatment of Wilson's disease, caused by an
accumulation of copper. As a consequence of its property of forming
stable complexes, it can also be employed to advantage for the elimination of other heavy metals (Pb, Hg) from the organism. In recent
years it has been increasingly utilized in connection with rheumatoid
arthritis [5] and neonatal jaundice [6]. In addition, Sorensen [7]
has demonstrated the antiinflammatory activity of the copper-D-
penicillamine complex.

Investigation of the metal complexes of L-cysteine and
D-penicillamine has a long past. In spite of this, their complex-
forming properties cannot be regarded as having been satisfactorily
clarified. As a result of the three complex-forming functional
groups (COO$^-$, -NH$_2$, and -S$^-$) L-cysteine and D-penicillamine are
ligands of an ambidentate nature. Since the -SH and -NH$_2$ groups
have relatively high pK values and the sulfur atoms may further
behave as bridging ligands, there are also many possibilities for
the formation of protonated and polynuclear complexes. Consequently,
the evaluation of the equilibrium conditions of their metal complexes
became possible only after the introduction of new computer methods.
However, the main difficulty in the study of the complexes of
L-cysteine and D-penicillamine lies in the possible redox reactions
of the mercapto group. If stable complexes may be formed between
the ligands and the metal ion in a number of different oxidation
states, then in general it is necessary to reckon with the parallel
occurrence of acid-base and redox reactions.

In this chapter we review the results of investigations connected with the possible complex formation processes of L-cysteine and D-penicillamine. In this work our starting point was the review by McAuliffe and Murray [8] which appeared in 1972; this contains an account of the results attained up to that time on the complexes of the sulfur-containing amino acids, and in part their critical analysis too. Accordingly, we deal here primarily with the results of more recent investigations. Earlier studies will receive attention only if the later results necessitate the reevaluation of the previous findings.

2. PROPERTIES OF THE LIGANDS

Both ligands contain three dissociable protons, with macroscopic protonation constants $pK_1 \sim 2$, $pK_2 \sim 8.0$, and $pK_3 \sim 10.5$. The value of pK_1 can be ascribed unambiguously to deprotonation of the carboxyl group, but the values of pK_2 and pK_3 arise from a combination of the microprocesses in Scheme 1 (where R = H (cysteine) and R = CH_3 (penicillamine); in the following, cys = cysteinate dianion, and pen = penicillaminate dianion).

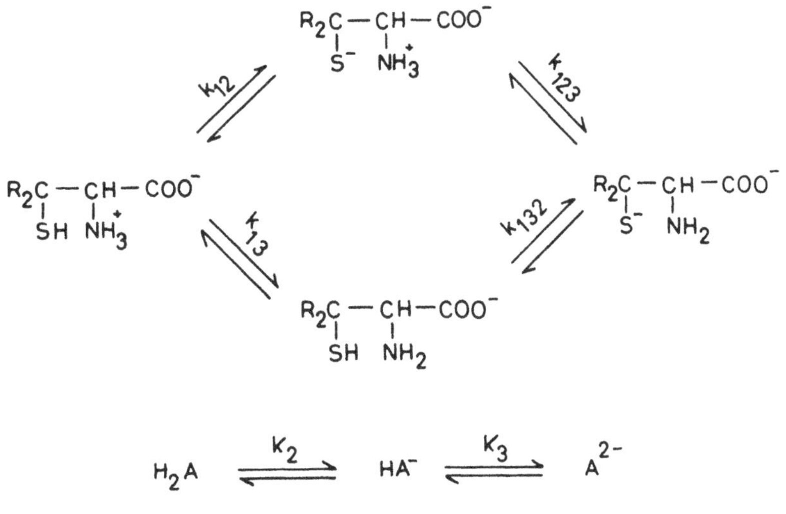

Scheme 1

Determination of the microconstants can be performed fairly easily by the joint application of pH metric and spectrophotometric [9-11] or nmr [12] methods. A comparison of data determined under identical conditions [13, 14] indicates that D-penicillamine has a slightly higher basicity because of the electron-repelling effect of the CH_3 groups. For steric reasons, the corresponding pK_2 and pK_3 values differ from one another to a more significant extent.

The metabolisms and biological roles of L-cysteine and D-penicillamine are in close connection with one of the most important reactions of the -SH group, i.e., oxidation. A possible pathway for this is illustrated in (1):

$$R\text{-}SH \xrightarrow{\text{ox}} R\text{-}S\text{-}S\text{-}R \xrightarrow{\text{ox}} R\text{-}SO_3H \tag{1}$$

In the presence of strong oxidants, the sulfonic acid is formed as the end product of the oxidation process. However, in the oxidation of every thiol derivative, an intermediate of outstanding importance and of marked stability with regard to further oxidation is the corresponding disulfide compound (cystine or D-penicillamine disulfide). From a coordination chemistry aspect it is very significant that (1) generally proceeds in the presence of catalysts [1]. Numerous experiments prove that the most effective catalysts are the various transition metal ions, and particularly copper(II) and iron(III). Hence, the study of the interactions of cysteine and D-penicillamine is necessary from the viewpoints of both complex formation and interpretation of the biological redox reactions.

The catalytic oxidation of D-penicillamine has not been studied to date, whereas the cysteine-cystine redox system has been investigated very extensively. Determination of the standard redox potential runs into difficulties, however, since the electrode processes are irreversible due to the formation of stable metal-cysteine complexes. Consequently, it is possible to determine the redox potentials of the thiol compounds only by indirect methods [1]. (According to the experiments of Jocelyn [15], E_0 (cysteine/cystine) = -0.22 V.)

Many publications deal with the interpretation of the mechanism of the metal ion-catalyzed oxidation of cysteine. Significant results have been attained primarily in connection with iron(III) [16, 17] and copper(II) [18, 19] ions. These references reveal that the catalysts are generally the transitionally formed various metal ion-cysteine complexes. Hanaki and Kamide [18] pointed out further that in the case of copper the active catalyst is the copper(II), which is present in comparatively constant concentration during the catalysis. The cause of this is that the reduction of the copper(II) ion with cysteine is a slower process than the reoxidation of the copper(I) complex with oxygen. At the same time, this may mean that in general more appreciable catalytic activities are exhibited by those metals that can form stable cysteine complexes in different valency states.

3. COMPLEXES OF NONTRANSITION ELEMENTS

3.1. Complexes of the Main Group Elements

With regard to the alkaline earth metal complexes, although certain contradictions are to be found among the earlier results [8, 20-22], investigations have not been performed in recent years.

Of the elements of main group III, results are primarily available in connection with gallium(III) and indium(III). Bianco et al. [23] assumed the formation of the complexes $Ga(cysH_2)^{3+}$, $Ga(cysH)^{2+}$, and $Ga(cys)^{+}$ in the gallium(III)-cysteine system in the interval $1 < pH < 3$. Kojima et al. [24] concluded that, besides the parent complexes $In(pen)^{+}$ and $In(pen)_2^{-}$, various protonated and hydroxo complexes, with compositions $In(penH)^{2+}$, $In(pen)_2H$, and $In(pen)(OH)$, are also likely to be formed in the indium(III)-DL-penicillamine system. The same authors suggest [25] that interpretation of the In(III)-cysteine equilibrium system necessitates assumption of the species $In(cys)_3^{3-}$ and $In(cysH)_2^{+}$ too. They have mentioned the formation of a sulfur-bridged polynuclear complex,

of the composition $In_3(cys)_4(OH)(H_2O)$. On the basis of their spectral examinations, they concluded that all three donor groups of the
ligands may participate in the coordination in the complexes InA^+
and InA_2^-. According to these authors, the protonated complexes contain S,O-bonded chelates, while the amino group remains protonated.

As regards the main group elements, the most wide-ranging
investigations have been made on the complexes of lead(II). As
already mentioned, D-penicillamine, together with other sulfur-
containing compounds, is extensively used in the treatment of lead
poisoning [26]. The early equilibrium examinations [20, 27] indicated that cysteine and penicillamine form very stable 1:1 chelates
with lead(II). Kuchinskas and Rosen [28] and Doornbos and Faber
[21] also assumed the formation of species of the composition MA_2
in the higher pH range. The latter authors [21] found that log
$K_1/K_2 \sim 9$. They explained this outstandingly high value in that all
three functional groups of the ligands take part in complex formation
in the 1:1 chelate, and thus the formation of MA_2 is sterically hindered. The more recent equilibrium [29] and thermochemical [30]
investigations of Corrie et al. likewise support the special, preferred role of the 1:1 chelate, but the formation of various protonated complexes was also suggested in lower concentration. In the
cases of both cysteine and D-penicillamine, however, all equilibrium
examinations exclude the possibility of the formation of polynuclear
complexes.

The tridentate nature of the ligands was confirmed by the ir
examinations of Shindo and Brown [31] and by the nmr examinations of
Natusch and Porter [32] and Sugiura et al. [11]. The complex Pb(pen)
was shown by Freeman et al. [33] in X-ray studies to have structure I.

In structure I, lead(II) is situated at the center of a distorted pentagonal bipyramid, in strong interaction with one N, one S,
and one O atom, and in weak interaction with two neighboring sulfur
atoms (S', S'') and one oxygen atom (O').

The investigations by Sugiura et al. [11] indicate that tin(II)
forms a complex similar to that of lead(II) with D-penicillamine.

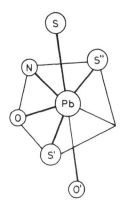

(I)

Their results appear to be confirmed by the more recent measurements of Gruenwedel and Hao [34], who reported the formation of 1:1 chelates for both L-cysteine and DL-penicillamine too.

In connection with the main group elements, the studies on the interaction of antimony(III) and D-penicillamine must also be mentioned. Pharmacological examinations prove that the harmful side effects of therapeutic preparations containing antimony(III) can be reduced significantly by the simultaneous administration of D-penicillamine [35]. Wandiga [35] carried out investigations with regard to the complexes formed. It proved possible to prepare the complex $K[Sb(pen)_2]$, but the nature of the bonding in this has not yet been established.

3.2. Complexes of Zinc, Cadmium, and Mercury

The complexes formed by zinc, cadmium, and mercury with sulfur-containing amino acids have been subjected to detailed study for a long time. This is due partly to the important biological role of zinc(II) and partly to the very toxic natures of cadmium(II) and mercury(II). Investigations relating to these systems were further enhanced considerably by the recognition that D-penicillamine can be very advantageously employed for the treatment of mercury poisoning; at the same time, it proved ineffective in the case of cadmium.

Most of the publications dealing with the zinc(II) complexes of cysteine and penicillamine [8, 12, 20, 21, 27, 28, 31, 32, 36] come to the general conclusion that only the complexes ZnA and ZnA_2^{2-} are formed: S,N-coordinated bidentate ligands were assumed in these complexes. The formation of protonated and polynuclear complexes was first pointed out by Perrin and Sayce [13]. Their results are supported by recent equilibrium data [14, 29]. Our own studies [14] suggested that polynuclear complexes are formed only in the zinc(II)-L-cysteine system, but the concentration of protonated complexes is also significant in the case of D-penicillamine. The concentration distributions of the species formed in the zinc(II)-L-cysteine and zinc(II)-D-penicillamine systems are depicted in Fig. 1.

The formation of the protonated complexes under the relatively high concentrations indicated in Fig. 1 proves that, besides S,N coordination, also S,O coordination plays an important role in the lower pH range. This finding is supported by the spectral studies of Zhegzda et al. [37] and Cothern et al. [38].

As regards the cadmium(II)-L-cysteine and cadmium(II)-D-penicillamine systems, Lenz and Martell [20] and later Sugiura et al. [11] came to the conclusion that complexes of the type MA are formed; cysteine and penicillamine act as tridentate ligands in these. Walker and Williams [39] also demonstrated formation of the complex $Cd(cys)_2^{2-}$ but obtained a very large value, 6.1, for $\log K_1/K_2$. In an X-ray study of the 1:1 Cd(II)-D-penicillamine complex, Freeman et al. [40] found that the cadmium(II) ion is hexacoordinated, being surrounded by two sulfur atoms, three oxygen atoms, and one nitrogen atom. Macroscopically, therefore, this complex has a polymeric structure. The bond distances are large compared to those in the corresponding lead(II) and mercury(II) complexes, which may explain the ineffectiveness of D-penicillamine in the biological elimination of cadmium. In addition to the above complexes, the species $Cd(cysH)_2$ and $Cd(penH)_2$ were recently prepared by Zhegzda et al. [41] and Carty and Taylor [42], respectively. In

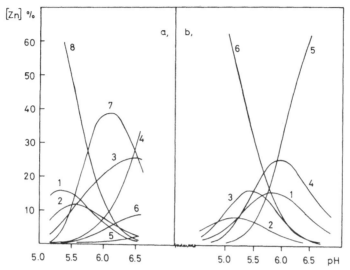

Fig. 1. Concentration-distribution of the complexes formed in zinc(II)-L-cysteine (a) or D-penicillamine (b) systems at 1:2 metal:ligand ratios. (a) 1--$Zn(cysH)^+$, 2--$Zn(cysH)_2$, 3--$Zn(cys)_2H^-$, 4--$Zn(cys)_2^{2-}$, 5--$Zn_2(cys)_3^{2-}$, 6--$Zn_3(cys)_4^{2-}$, 7--$Zn_3(cys)_4H^-$, 8--Zn^{2+}. (b) 1--$Zn(pen)$, 2--$Zn(penH)^+$, 3--$Zn(penH)_2$, 4--$Zn(pen)_2H^-$, 5--$Zn(pen)_2^{2-}$, 6--Zn^{2+}.

these compounds, similarly as for zinc(II), the amino group remains protonated, i.e., S,O coordination may occur in the case of cadmium(II) as well.

As a consequence of the fairly simple experimental procedures, the results of the large number of investigations carried out since the turn of the century with regard to mercury(II) were for a long time fairly contradictory [8]. On the basis of the more recent studies, however, the compositions and structures of the very stable mercury(II)-cysteine and mercury(II)-D-penicillamine complexes can already be given to a good approximation. Van der Linden and Beers [43] determined the stability constant of the complex $Hg(cysH)_2$ and concluded that cysteine is bonded to the mercury(II) in a monodentate manner via the sulfur atom, the amino group remaining protonated. At higher pH, in the case of a mercury(II) excess, however, the possibility arises for the formation of polynuclear complexes

containing Hg-N bonds too. On the basis of the spectral data of
$[Hg(cysH)_2]HCl \cdot 1/2H_2O$, Neville and Drakenberg [44] similarly conclude
that the Hg-S bonding plays a preferred role. By means of X-ray
investigations, Carty and Taylor [45] proved the structures of the
DL-penicillamine complexes $(HgCl_2)_2(penH_2) \cdot 2H_2O$ and $HgCl_2(penH_2)_2 \cdot H_2O$,
and pointed out the important roles played by the chlorine and sulfur
donor atoms in the formation of the structures.

Recently an increasing number of publications are dealing with
the interactions of sulfur-containing amino acids and organomercury
compounds, which are extremely strong poisons and cause severe cere-
bral damage [46-50]. These studies lead to the uniform result that
methyl mercury (and other derivatives) form very stable 1:1 monomeric
complexes with cysteine and penicillamine, in which exclusively Hg-S
bonding occurs. In the event of an organomercury excess, however,
compounds of the composition $(RHg)_2(pen)$ may also be formed in which
the second mercury atom is bonded via the amino group.

4. COMPLEXES OF THE TRANSITION METALS

In this section the discussion will be restricted only to the complexes of those metal ions that have been studied most extensively
and that are also important from a biological aspect (chromium,
molybdenum, iron, cobalt, nickel, copper, and the platinum metals).
As to the other complexes we refer again to the review by McAuliffe
and Murray [8].

4.1. Complexes of Chromium and Molybdenum

Only in the most recent investigations did it become certain that
the chromium(III) ion plays a basic role in biochemical processes
[51]. It has long been known, however, that both chromium(III) and
chromium(VI) compounds are very toxic. Chromium(VI) compounds in
particular may cause extremely severe damage in the human organism.

As a consequence of the presumed role of the Cr-S bond in biochemical processes, investigations have recently been commenced to throw light on the interactions between chromium(III) and chromium(VI) and the sulfur-containing amino acids.

The chromium-D-penicillamine system has been studied in detail by Sugiura et al. [51, 52], who found that during the interaction of chromium(III) and D-penicillamine a chromium(III) complex of the composition $Cr(pen)_2^-$ is formed in a slow reaction. If CrO_4^{2-} and D-penicillamine were reacted together, formation of the same product in a relatively fast process was detected. The occurrence of the following processes was assumed in this reaction:

$$Cr(VI) + H_3pen \longrightarrow Cr(V) + disulfide \qquad (2)$$

$$Cr(V) + 2H_3pen \longrightarrow [Cr(V)(pen)_2]^+ \qquad (3)$$

$$[Cr(V)(pen)_2]^+ \xrightarrow[\text{by solvent}]{\text{reduction}} [Cr(III)(pen)_2]^- \qquad (4)$$

Sugiura et al. [51, 52] concluded that the ligand has a tridentate nature in the complex $[Cr(pen)_2]^-$. This was supported by de Meester and Hodgson [53], who prepared the mixed ligand complex Cr(III)(L-histidine)(D-pen) and demonstrated the tridentate nature of the D-penicillamine by X-ray examinations.

Studies aimed at determining the bonding sites of molybdenum-containing enzymes have shown that the Mo-S bond has a preferred role. Hence, numerous publications have dealt with investigations of the molybdenum-cysteine and molybdenum-penicillamine systems. The central features of these investigations are the study of the complicated redox and complex formation processes in the interaction of molybdenum(VI) and the sulfur-containing amino acids, and the establishment of which of these compounds are suitable for the modeling of the catalytic reactions of molybdenum-containing enzymes [54].

As regards the MoO_4^{2-}-cysteine system, it was first reported by Spence and Chang [55] that molybdenum(V) complexes with compositions of 1:1 and molybdenum(VI) complexes with compositions of 1:1, 1:2, and 1:3 may be formed. On the other hand, Martin and Spence

[56] demonstrated that an interaction between molybdenum(VI) and cysteine cannot be proved, and only the molybdenum(V) produced as a result of the redox reaction forms a complex with the excess cysteine. By X-ray examination, Knox and Prout [57] found structure II for the molybdenum(V) $[Mo_2O_4(cys)_2]^{2-}$ 1:1 complex.

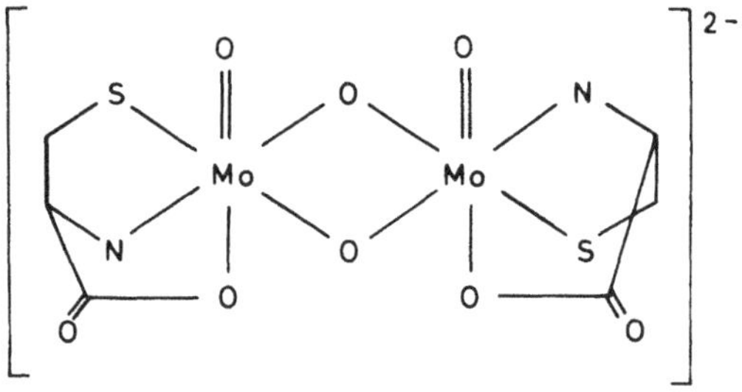

(II)

Structure II shows that although the compound contains tridentate cysteine, the carboxyl group is more weakly bound to the metal. It was shown by Huang and Haight [58] that, with the increase of the pH, the yellow, diamagnetic complex with structure II is converted into a deeper-colored esr-active compound, for which the monomeric structure III was suggested.

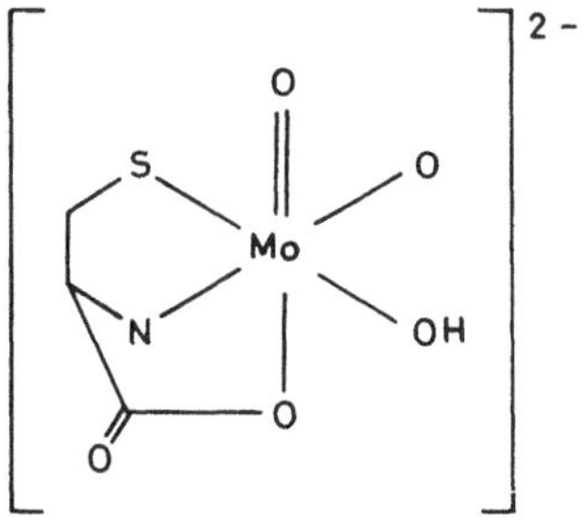

(III)

On the other hand, Kroneck and Spence [59] interpret the transformation of structure II in alkaline medium by the formation of an esr-active dimeric complex with structure IV. In addition, it was pointed out by Kay and Mitchell [60] that on the action of H_2S complex II can be converted into a sulfide-bridged form which is always diamagnetic.

(IV)

The catalytic activity of complex II as a biological model has
been extensively studied by Schrauzer et al. [61]. They concluded
that a transitionally formed molybdenum(IV)-cysteine complex is the
catalytically active species. More recently, Lamache-Duhameaux [62]
demonstrated the existence of a molybdenum(IV)-cysteine complex in
solution. In contradiction to the former conclusions on the basis
of their electrochemical examinations, Ott and Schultz [63] attribute
catalytic activity to the molybdenum(III)-cysteine complex.

The interaction between molybdenum and D-penicillamine has been
studied in detail by Sugiura et al. [64]. Their results indicate
that the reactions involved are similar to those of cysteine, but
the formation of a 1:2 molybdenum(III)-D-penicillamine complex was
also suggested.

4.2. Complexes of Iron and Cobalt

The nonheme iron proteins in general contain Fe-S bonds, and therefore the iron(II) and iron(III) complexes have been dealt with for
a long time [8]. However, the study of the complexes formed is
fairly complicated because of the redox processes. If iron(II)
reacts with cysteine, the complexes formed are very easily oxidized.
On the other hand, the oxidation of cysteine is catalyzed by the
iron(III) ion, and hence the total cysteine can be relatively
rapidly transformed to cystine. The short life times of the complexes make their study very difficult. Efforts to solve the early
contradictory results were made by Tomita et al. [65], who demonstrated that the short-lived species too can be well examined in

frozen solutions. They described the formation of four different
iron(III) complexes.

(a)   (b)

(c)   (d)

(v)

The 1:1 blue complex Va is formed only in aqueous acidic
medium. In alkaline solution, formation of the intensely red Vc
and violet Vb is favored. The green tris complex Vd is a coordination isomer of the violet compound. According to Tomita et al., in
all cases the cysteine acts as a bidentate ligand in S,N or S,O
coordination, and the products are monomers. Bell et al. [66],
however, claim that polymeric complexes may be formed instead of
those with structures Va-Vd. This assumption is supported by
Murray and Newman [67], who prepared two water-insoluble iron(II)-
cysteine complexes that are unstable in air, with stoichiometric
compositions of 1:1 (structure VI) and 1:2 (structure VII). Their
magnetic and spectral properties indicated that the complexes are
polymers.

(VI)   (VII)

In connection with the iron(III)-D-penicillamine system, Bell et al. [66] found that at least two intensely-colored (red and violet) iron(III) complexes exist, for which they suggested a stoichiometric composition of 1:2. They also showed that formation of the colored complexes is much more marked in this system, and in contrast with those of cysteine the complexes can be studied at room temperature. According to the work of Stadtherr and Martin [68], the red compound is a bis complex of the composition $Fe(pen)_2^-$ that can be obtained in alkaline medium either in the reaction iron(II) + D-penicillamine + oxidant, or in the direct reaction iron(III) + D-penicillamine. In contrast with the previous authors, however, from the absence of stereoselectivity as indicated by CD measurement, the penicillamine was concluded to be tridentate. They further described the formation of a blue iron(III) complex analogous to structure Va, and an iron(II) complex, $Fe(pen)_2^{2-}$.

The significant results on the cobalt(II)-cysteine and cobalt(III)-cysteine systems were attained earlier by Schubert [69] and Neville [70a] and these are discussed in detail by McAuliffe and Murray [8]. A comparison of the results reveals that both the cobalt(II) and the cobalt(III) ion form very stable bis and tris complexes with cysteine. The cobalt(II) compounds, however, are readily oxidized in air. In these complexes the cysteine generally acts as an S,N bidentate ligand, although in the case of $Co(cys)_3^{3-}$

S,O coordination is also possible. In addition to the mononuclear
complexes the formation of a polimeric species of $Co_3A_4^{2-}$ was proved
by Williams et al. [70b].

The interaction between cobalt(II) and D-penicillamine has
recently been investigated by Boggess and Martin [71]. These authors
came to the conclusion that only a bis complex of the composition
$Co(pen)_2^{2-}$ is formed; although the carboxyl group in this can readily
be replaced by a water molecule, the ligand is nevertheless tridentate. It is surprising that very little attention has been paid to
equilibrium studies of these systems [12, 20]. At any event, the
existence of coordination isomers makes it possible for protonated
and complexes to form in this system too, similarly as for zinc(II)
and nickel(II). Consequently, further equilibrium studies of these
systems is justified.

4.3. Complexes of Nickel, Palladium, and Platinum

A number of authors have studied the complexes of nickel(II) with
L-cysteine and D-penicillamine from both equilibrium and structural
aspects. In the early measurements, however, only the formation of
the parent complexes NiA and NiA_2^{2-} was assumed [12, 20, 21, 28, 72,
73]. In addition, it was pointed out that S,N coordinated, diamagnetic 1:2 complexes with square-planar geometry are formed [72, 74].
Attention was drawn to the formation of protonated and polynuclear
complexes by Perrin and Sayce [13], and this was supported by the
recent investigations [14]. Protonated complexes may also be formed
in low concentration in the Ni(II)-D-penicillamine system, but the
polynuclear complex of the composition $M_3A_4^{2-}$ (structure VIII) is
primarily characteristic of the Ni(II)-L-cysteine system. In the
case of nickel(II)-D-penicillamine, the failure of the polymer to
form is just explained by the assumption of structure VIII. (Because
of the steric hindrance caused by the presence of the CH_3 groups a
polymer complex cannot be formed.)

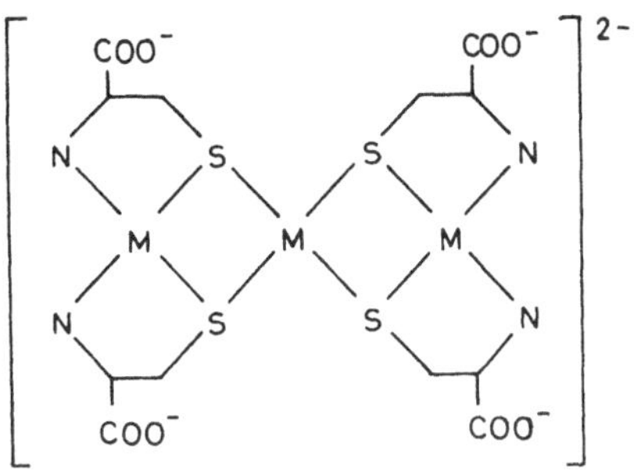

(VIII)

The results of McAuliffe et al. [36, 75] showed that protonated and polynuclear complexes may also be formed in the case of Pd(II). They prepared compounds with the compositions $[Pd_2(cysH)_3Cl] \cdot 2H_2O$, $[Pd(cysH)Cl]_2$, and $[Pd(penH)Cl]_2$, and proposed structure IX for the latter two.

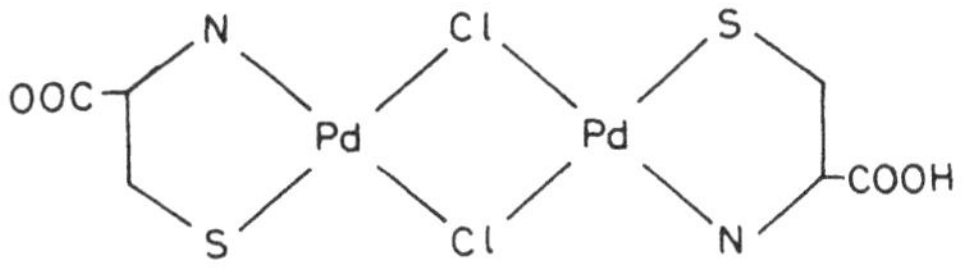

(IX)

Palladium(II) and platinum(II) complexes of cysteine were prepared by Chandrasekharan et al. [76] too, but in contrast with the former authors they suggested S,O coordination for the square-planar compounds. On the other hand, Volstein and Krilova [77] prepared the complex $[Pt(cysH)Cl]_2$, analogous to structure IX, and $Pt(cysH_2)Cl_2$, involving coordination only by sulfur atoms.

4.4. Complexes of Copper

Reference has already been made to the therapeutic and biochemical significance of the interaction between copper(II) and D-penicillamine, and to the role of copper(II) in the oxidation of cysteine. The study of systems containing copper(II) and a thiol compound is appreciably

hampered by the fact that redox and complex formation processes occur
in parallel, while both the metal ion and the ligand may form complexes that are stable in either the oxidized or the reduced form.

Attention was first drawn by Vallon and Badinand [78] to the
formation of a 1:1 Cu(I)-D-penicillamine complex in the event of a
D-penicillamine excess in the copper(II)-D-penicillamine system.
According to the recent investigations by Gergely and Sóvágó [79],
however, a polymer with a stoichiometry $Na[Cu(I)_n(penH)_{n+1}]$ is
obtained in this process.

Sugiura and Tanaka [80] and Wilson and Martin [81] were the
first to point out that at a definite metal-ligand ratio an intense
red-violet complex is formed in the copper(II)-D-penicillamine
system. It was assumed that both copper(II) and copper(I) play a
role in this compound. After studying the optimum conditions for
formation of the compound, Wright and Frieden [82] and Musker and
Neagley [83] came to the conclusion that the mixed valence state is
stabilized only in the presence of Cl^- or Br^-. These authors also prepared the red-violet complex in the solid state and put forward various
polymeric structures for it. On the basis of their X-ray examinations,
Birker and Freeman [84] recently proposed a cluster structure for the
complex anion of composition $[Cu(I)_8Cu(II)_6(pen)_{12}Cl]^{5-}$. In this,
each of the six copper(II) atoms is surrounded by two D-penicillamine
molecules in S,N coordination. Additionally, the sulfur atoms are
also bonded to the copper(I) ions, and the bonding system $Cu(I)_8S_{12}$
forms the skeleton of the cluster. At the same time it was pointed
out that the existence of this unusual valence state can be explained
primarily by the role of the β substituents, which is in agreement
with the earlier findings of Sugiura and Tanaka [80]. The above
structure was supported by the investigations of Schugar et al. [85].

The formation conditions and reactions of the mixed valence
complex have been studied in detail by Gergely and Sóvágó [79].
They found that the mixed valence state is stabilized only at a well-
defined halide concentration, as illustrated in Fig. 2.

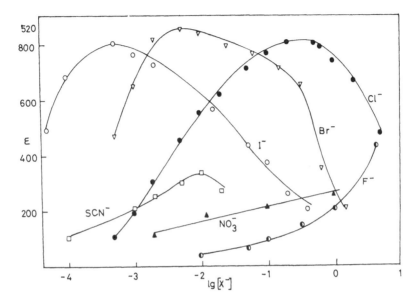

FIG. 2. Variation of the values of molar absorption coefficients
calculated for the total copper content in the systems containing
copper(II) and D-penicillamine at the ratio 1:1.45.

Their magnetic measurements indicate that the magnetic moment
of the red-violet complex corresponds to 44% of the total copper
content. It was further demonstrated in agreement with the findings
of Birker and Freeman [84] that the compound is esr-active, and the
solid possesses a slight electric conductivity [79]. It emerged
from a study of the reactions of the mixed valence complex that in
acidic medium in the presence of halide ion it decomposes via formation of a short-lived, intense-blue complex. This product can also
be detected in acidic medium in systems with a copper(II)-D-penicill-
amine composition of 1:2, and its spectral properties are reminiscent
of those of the blue copper proteins [3].

    In contrast, Laurie et al. [86] and Rupp and Weser [87] reject
the existence of the mixed valence state. At the same time, Laurie
et al. [86] point out that in the presence of glycylglycine or histidine the redox reaction between D-penicillamine and the copper(II)
ion is repressed, and a copper(II) mixed ligand complex is formed.

Studies on the copper(II)-cysteine interaction are much less
extensive. Although Stricks and Kolthoff [88] described the formation of a copper(I)-cysteine polymer in 1951, since that time it has
not proved possible to prepare other stable complexes. Hawkins and
Perrin [89] investigated the formation of the copper(II)-cystine
complex, which is analogous to the dimeric copper(II) complex of
D-penicillamine disulfide, and in which the ligand participates via
N,O coordination [90]. Hanaki [91] recently observed the formation
of a short-lived, intense-purple copper cysteine complex too.
Electron spin resonance examinations indicate formation of a
copper(II)-cysteine complex, but this is very rapidly converted
to a copper(I) complex.

A detailed study of the copper(II)-L-cysteine system was made
by us [92]. It was found that the analogous mixed valence complex
is also obtained in the case of cysteine, at a KI concentration of
2 M. However, the lifetime of the product in aqueous solution is
substantially shorter than that of the corresponding D-penicillamine
complex, presumably because of the greater hydration of the copper(I)
ion [84].

It is clear from the foregoing that the copper(II)-cysteine
and copper(II)-D-penicillamine interactions have not been at all well
clarified in many of their details. The mechanism of the biological
elimination of copper is similarly uncertain, as no clear proof is
available as to whether the copper(I) complex [78], the mixed ligand
copper(II) complex [86], or the mixed valence complex [82, 93, 94]
plays a role in this respect.

5. MIXED LIGAND COMPLEXES

It was first pointed out by Sugiura and Tanaka [95] that 1:1:1 mixed
ligand complexes are formed in solutions containing lead(II),
cadmium(II), or mercury(II) and D-penicillamine, together with other
sulfur-containing ligands. In the case of glycine or histidine as
B ligands, however, mixed ligand complex formation was not observed.

Chang and Martin [74] reached similar findings from a study of the
nickel(II)-cysteine-ethylenediamine system. This is in agreement
with the experimental observation (generally regarded as true) that
mixed ligand nickel(II) complexes are not formed if one of the
parent complexes is square-planar and the other octahedral [96].
Nevertheless, in the nickel(II)-D-penicillamine-histidine (or histamine) systems [14] the formation of a small amount of mixed ligand
complex was observed, which could be well identified; this was interpreted by the occurrence of a square-planar $\rightleftharpoons$ octahedral equilibrium in solution. Mixed ligand complex formation in the zinc(II)
complexes of similar composition was found to be close to the
statistical case.

De Meester and Hodgson [53, 97] prepared mixed ligand complexes
similarly containing L-histidine and D-penicillamine in the cases of
chromium(III) and cobalt(III). The ligands in these complexes are
tridentate.

The mixed ligand complexes of the copper(II) ion are particularly interesting [79, 86, 91, 93]. So far, the preparation of a
D-penicillamine complex of the type CuA_2 containing the copper(II)
ion has not succeeded because of redox reactions. However, a stable
copper(II) ion is present in the mixed ligand complexes copper(II)-
D-penicillamine-L-histidine or glycylglycine.

6. CONCLUSIONS

The ambidentate nature of the ligands L-cysteine and D-penicillamine
may be manifested in the formation of chelate rings involving various
coordinations (S,N; S,O; or N,O), and also in the number of coordination sites occupied by the ligands.

In the lead(II), cadmium(II), chromium(III), and molybdenum(V)
complexes, L-cysteine and D-penicillamine participate in complex
formation with all three donor groups in general. Copper(I),
mercury(II), and silver(I) do not generally form chelates and enter
into coordinate bonding only with the sulfur atom. In the case of

a metal ion excess, polynuclear complexes too may be obtained via
the sulfur atom and possibly the $-NH_2$ group. The transition metals
iron(II), iron(III), cobalt(II), cobalt(III), nickel(II), palladium(II)
platinum(II), copper(II), and zinc(II) usually form chelates with
L-cysteine and D-penicillamine, and bidentate, or in certain cases
tridentate, coordination of the ligand is assumed. In most cases
coordination occurs via the S and N donor atoms, but S,O coordination
too may be found; this is primarily characteristic of iron(III), but
also to some extent of cobalt(III) and zinc(II).

The behavior is particularly complicated in those systems where
redox processes also may ensue during the metal ion-ligand interaction
(chromium, molybdenum, iron, cobalt, and copper). L-cysteine and
D-penicillamine are soft-character ligands, and generally form thermo-
dynamically more stable complexes with metal ions in lower valence
states. However, these compounds are usually very readily oxidized
[e.g., iron(II)-cysteine, copper(I)-D-penicillamine], i.e., the metal
ions act as catalysts for the oxidation of the ligands.

Comparison of the available experimental results allows us to
point out some of the more important differences between the complex-
forming properties of L-cysteine and D-penicillamine. It may be
stated that the formation of polynuclear complexes in which the
sulfur atom acts as the bridging ligand is primarily characteristic
of cysteine [at least when complexed to indium(III), nickel(II), and
zinc(II)]. As regards the parallel acid-base and redox reactions,
the reducing effect of D-penicillamine is smaller, and therefore the
higher or mixed valence state complexes are more stable. Thus, compared to L-cysteine, D-penicillamine is a better model compound for
studying the bonding sites of the various metalloproteins.

REFERENCES

1. P. C. Jocelyn, Biochemistry of the SH group, Academic Press,
    London and New York, 1972.

2. G. R. Moore and R. J. P. Williams, *Coord. Chem. Rev.*, *18*, 125
    (1976).

3. J. A. Fee, "Copper Proteins" in Structure and Bonding, Vol. 23, Springer-Verlag Berlin, 1975.

4. O. Sazukin and S. M. Navarin, *Antibiotiki*, *6*, 562 (1965).

5. D. Perrett, W. Sneddon, and A. D. Stephens, *Biochem. Pharmacol.*, *25*, 259 (1976).

6. L. Lakatos, B. Kövér, Gy. Oroszlán, and Zs. Vekerdy, *Eur. J. Pediat.*, *123*, 133 (1976).

7. J. R. Sorensen, *J. Med. Chem.*, *19*, 135 (1976).

8. C. A. McAuliffe and S. G. Murray, *Inorg. Chim. Acta Reviews*, *6*, 103 (1972).

9. E. Coates, C. Marsden and B. Rigg, *Trans. Faraday Soc.*, *65*, 863 and 3032 (1969).

10. G. E. Clement and T. P. Hartz, *J. Chem. Educ.*, *48*, 395 (1971).

11. Y. Sugiura, A. Yokoyjama, and H. Tanaka, *Chem. Pharm. Bull.*, *18*, 693 (1970).

12. D. B. Walters and D. E. Leyden, *Anal. Chim. Acta*, *72*, 275 (1974).

13. D. D. Perrin and I. G. Sayce, *J. Chem. Soc. A.*, 53 (1968).

14. I. Sóvágó, A. Gergely, B. Harman, and T. Kiss, *J. Inorg. Nucl. Chem.* (in press).

15. P. C. Jocelyn, *European J. Biochem.*, *2*, 327 (1967).

16. N. Tanaka, I. M. Kolthoff, and W. Stricks, *J. Amer. Chem. Soc.*, *77*, 2004 (1955).

17. J. A. Taylor, J. F. Yan and J. Wang, *J. Amer. Chem. Soc.*, *88*, 1663 (1966).

18. A. Hanaki and H. Kamide, *Chem. Pharm. Bull.*, *19*, 1006 (1971) and *23*, 1671 (1975).

19. G. J. Bridgart, M. W. Fuller, and I. R. Wilson, *J. Chem. Soc. Dalton*, 1274 (1973).

20. G. R. Lenz and A. E. Martell, *Biochemistry*, *3*, 745 (1964).

21. D. A. Doornbos and J. S. Faber, *Pharm. Weekblad*, *99*, 289 (1964).

22. T. Tang, K. S. Rajan, and N. Greez, *Biophys. J.*, *8*, 1458 (1968).

23. P. Bianco, J. Haladjian, and R. Pilard, *J. Electroanal. Chem.*, *72*, 341 (1976).

24. N. Kojima, Y. Sugiura, and H. Tanaka, *Bull. Chem. Soc. Japan*, *49*, 1294 (1976).

25. N. Kojima, Y. Sugiura, and H. Tanaka, *Bull. Chem. Soc. Japan*, *49*, 3023 (1976).

26. D. D. Perrin, "Inorganic Medicinal Chemistry," in Topics in Current Chemistry, Vol. 52, Springer-Verlag, Berlin-Heidelberg-New York, 1974.

27. N. C. Li and R. A. Manning, *J. Amer. Chem. Soc.*, *77*, 5225 (1955).

28. E. J. Kuchinskas and Y. Rosen, *Arch. Biochem. Biophys.*, *97*, 370 (1962).

29. A. M. Corrie, M. D. Walker, and D. R. Williams, *J. Chem. Soc. Dalton*, 1012 (1976).

30. A. M. Corrie and D. R. Williams, *J. Chem. Soc. Dalton*, 1068 (1976).

31. H. Shindo and T. L. Brown, *J. Amer. Chem. Soc.*, *87*, 1904 (1965).

32. D. F. S. Natusch and L. J. Porter, *J. Chem. Soc. A.*, 2527 (1971).

33. H. C. Freeman, G. N. Stevens, and I. F. Taylor, *J. Chem. Soc., Chem. Comm.*, 366 (1974).

34. D. W. Gruenwedel and H.-C. Hao, *J. Agr. Food. Chem.*, *21*, 246 (1973).

35. S. O. Wandiga, *J. Chem. Soc. Dalton*, 1894 (1975).

36. S. T. Chow, C. A. McAuliffe, and B. J. Sayle, *J. Inorg. Nucl. Chem.*, *35*, 4349 (1973).

37. G. D. Zhegzda, A. P. Gulia, S. I. Neikowskii, and F. M. Tulupa, *Khoord. Khim.*, *2*, 1031 (1976).

38. C. R. Cothern, W. E. Moddeman, R. G. Albridge, W. J. Sanders, P. L. Kelly, W. S. Hanley, and L. Field, *Anal. Chem.*, *48*, 162 (1976).

39. M. D. Walker and D. R. Williams, *J. Chem. Soc. Dalton*, 1186 (1974).

40. H. C. Freeman, F. Huq, and G. N. Stevens, *J. Chem. Soc., Chem. Comm.*, 90 (1976).

41. G. D. Zhegzda, V. N. Kabanova, and F. M. Tulupa, *Zh. Neorg. Khim.*, *20*, 2325 (1975).

42. A. J. Carty and N. J. Taylor, *Inorg. Chem.*, *16*, 177 (1977).

43. W. E. Van der Linden and C. Beers, *Anal. Chim. Acta*, *68*, 143 (1973).

44. G. A. Neville and T. Drakenberg, *Can. J. Chem.*, *52*, 616 (1974).

45. A. J. Carty and N. J. Taylor, *J. Chem. Soc., Chem. Comm.*, 214 (1976).

46. Y. S. Wong, P. C. Chieh, and A. J. Carty, *Can. J. Chem.*, *51*, 2597 (1973) and *J. Chem. Soc., Chem. Comm.*, 741 (1973).

47. Y. S. Wong, N. J. Taylor, P. C. Chieh, and A. J. Carty, *J. Chem. Soc., Chem. Comm.*, 625 (1974).

48. N. J. Taylor, Y. S. Wong, P. C. Chieh, and A. J. Carty, *J. Chem. Soc. Dalton*, 438 (1975).

49. D. L. Rabenstein and M. T. Fairhurst, *J. Amer. Chem. Soc.*, *97*, 2086 (1975).

50. Y. Hojo, Y. Sugiura, and H. Tanaka, *J. Inorg. Nucl. Chem.*, *38*, 641 (1976).

51. Y. Hojo, Y. Sugiura, and H. Tanaka, *J. Inorg. Nucl. Chem.*, *39*, 1859 (1977).

52. Y. Sugiura, Y. Hojo, and H. Tanaka, *Chem. Pharm. Bull.*, *20*, 1362 (1972).

53. P. de Meester and D. J. Hodgson, *J. Chem. Soc. Dalton*, 1604 (1977) and *J. Chem. Soc., Chem. Comm.*, 280 (1976).

54. J. T. Spence, *Coord. Chem. Rev.*, *4*, 475 (1969).

55. J. T. Spence and H. H. Y. Chang, *Inorg. Chem.*, *2*, 319 (1963).

56. J. F. Martin and J. T. Spence, *J. Phys. Chem.*, *74*, 2863 (1970).

57. J. R. Knox and C. K. Prout, *J. Chem. Soc., Chem. Comm.*, 1227 (1968).

58. T. J. Huang and G. P. Haight Jr., *J. Amer. Chem. Soc.*, *92*, 2336 (1970).

59. P. Kroneck and J. T. Spence, *Inorg. Nucl. Chem. Letters*, *9*, 177 (1973); and *J. Inorg. Nucl. Chem.*, *35*, 3391 (1973).

60. A. Kay and P. C. H. Mitchell, *J. Chem. Soc. A.*, 2421 (1970).

61. G. N. Schrauzer, G. W. Kiefer, K. Tano, and P. A. Doemeny, *J. Amer. Chem. Soc.*, *96*, 641 (1974); other references see here.

62. M. Lamache-Duhameaux, *J. Inorg. Nucl. Chem.*, *39*, 2081 (1977).

63. V. R. Ott and F. A. Schultz, *Electroanal. Chem. Int. Electrochem.*, *61*, 81 (1975).

64. Y. Sugiura, T. Kikuchi, and H. Tanaka, *Chem. Pharm. Bull.*, *25*, 345 (1977).

65. A. Tomita, H. Hirai, and S. Makashima, *Inorg. Chem.*, *6*, 1746 (1967); ibid., *7*, 760 (1968).

66. C. M. Bell, E. D. McKenzie, and J. Orton, *Inorg. Chim. Acta*, *5*, 109 (1971).

67. K. S. Murray and P. J. Newman, *Aust. J. Chem.*, *28*, 773 (1975).

68. L. G. Stadtherr and R. B. Martin, *Inorg. Chem.*, *11*, 92 (1972).

69. M. P. Schubert, *J. Amer. Chem. Soc.*, *55*, 3336 (1933).

70. (a) R. G. Neville, *J. Amer. Chem. Soc.*, *79*, 518 (1957),
    (b) K. Garbet, G. W. Partridge, and R. J. P. Williams, *Bioinorg. Chem.*, *1*, 309 (1972).

71. R. K. Boggess and R. B. Martin, *J. Inorg. Nucl. Chem.*, *37*, 359 (1975).

72. S. K. Srivastava, E. V. Raju, and H. B. Mathur, *J. Inorg. Nucl. Chem.*, *35*, 253 (1973).

73. J. H. Ritsma and F. Jellinek, *Recueil*, *91*, 923 (1972).

74. J. W. Chang and R. B. Martin, *J. Phys. Chem., 73,* 4277 (1969).

75. W. Levason, C. A. McAuliffe, and D. M. Johns, *Inorg. Nucl. Chem. Lett., 13,* 123 (1977).

76. M. Chandrasekharan, M. R. Udupa, and G. Aravamudan, *Inorg. Chim. Acta, 7,* 88 (1973).

77. L. M. Volstein and L. F. Krilova, *Zh. Neorg. Khim., 21,* 2250 (1976).

78. J. J. Vallon and A. Badinand, *Anal. Chim. Acta, 42,* 445 (1968).

79. A. Gergely and I. Sóvágó, *Bioinorg. Chem., 9,* 47 (1978).

80. Y. Sugiura and H. Tanaka, *Chem. Pharm. Bull., 18,* 368 (1970).

81. E. W. Wilson and R. B. Martin, *Arch. Biochem. Biophys., 142,* 445 (1971).

82. J. R. Wright and E. Frieden, *Bioinorg. Chem., 4,* 163 (1975).

83. W. K. Musker and C. H. Neagley, *Inorg. Chem., 14,* 1728 (1975).

84. P. J. M. W. L. Birker and H. C. Freeman, *J. Chem. Soc., Chem. Comm.,* 312 (1976) and *J. Amer. Chem. Soc., 99,* 6890 (1977).

85. H. J. Schugar, C. Ou, J. A. Thich, J. A. Potenza, R. A. Lalancette and W. Furey Jr., *J. Amer. Chem. Soc., 98,* 3047 (1976).

86. S. H. Laurie, T. Lund, and J. B. Raynor, *J. Chem. Soc. Dalton,* 1389 (1975).

87. H. Rupp and U. Weser, *Biochim. Biophys. Acta, 446,* 151 (1976).

88. W. Stricks and I. M. Kolthoff, *J. Amer. Chem. Soc., 73,* 1723 (1951).

89. C. J. Hawkins and D. D. Perrin, *Inorg. Chem., 2,* 843 (1963).

90. J. A. Thich, D. Mastropaolo, J. Potenza, and H. J. Schugar, *J. Amer. Chem. Soc., 96,* 726 (1974).

91. A. Hanaki, *Chem. Pharm. Bull., 21,* 2491 (1974) and *Chem. Lett.,* 1225 (1976).

92. I. Sóvágó, B. Harman, and A. Gergely, in preparation.

93. Y. Sugiura and H. Tanaka, *Mol. Pharmacol., 8,* 249 (1972).

94. T. Shalouhi, P. T. Evans, and J. R. Wright, *Physiol. Chem. Phys., 8,* 337 (1976).

95. Y. Sugiura and H. Tanaka, *Chem. Pharm. Bull., 18,* 746 (1970).

96. R. P. Martin, M. M. Petit-Ramel, and J. P. Scharff, "Mixed-Ligand Metal Ion Complexes of Amino Acids and Peptides" in Metal Ions in Biological Systems (H. Sigel, ed.), Marcel Dekker, New York, Vol. 2, 1973, p. 1.

97. P. de Meester and D. J. Hodgson, *J. Amer. Chem. Soc., 99,* 101 (1977).

Chapter 4

GLUTATHIONE AND ITS METAL COMPLEXES

Dallas L. Rabenstein, Roger Guevremont,
and Christopher A. Evans
Department of Chemistry
The University of Alberta
Edmonton, Alberta, Canada

1. INTRODUCTION

1.1. General

Glutathione (GSH) is a naturally occurring tripeptide with the
sequence γ-L-glutamyl-L-cysteinyl-glycine. The structural formula
of the predominant isomer at neutral pH is

$$^-O_2CCHCH_2CH_2\overset{\overset{O}{\|}}{C}NHCHCNHCH_2CO_2^-$$

In its reactions with metal ions, GSH presents eight potential
binding sites: two carboxylic acid groups, an amino group, a
sulfhydryl group, and two peptide linkages. Because all eight
binding sites cannot be simultaneously coordinated to a single metal
ion, the coordination chemistry of GSH is characterized by the formation of protonated and polynuclear complexes. The complexes of GSH
with some metals, e.g., molybdenum and copper, have been studied as
models for their behavior in more complicated biological systems.
The coordination chemistry of GSH is also of interest in its own
right.

     Much of the emphasis in studies with labile metal ions has been
on characterizing the nature of the complexes and their equilibria in

aqueous solution. In these studies, potentiometric methods have been used most frequently and, as more sophisticated computer-based procedures for extracting formation constants from potentiometric titration measurements have been developed, more detailed models have been developed for the complexation equilibria. However, potentiometric measurements alone provide little or no information about the nature of the complexes at the molecular level, and the validity of the results from these studies is dependent on the correctness of the models deduced in fitting the experimental data. In several instances, quite different models have been derived from potentiometric titration data. These problems have been resolved to some extent by making use of information about the nature of the complexes at the molecular level, obtained for example from spectroscopic measurements.

In this chapter, we review the aqueous solution chemistry of GSH and its metal complexes. Complexes of several metals with the disulfide molecule derived from GSH by oxidation of its sulfhydryl group (GSSG) have also been studied, and these are discussed. The emphasis in this review is on the nature of the complexes at the molecular level, and attempts are made to resolve those cases where conflicting interpretations have been presented.

1.2. Biological Occurrence of Glutathione

Following its crystallization from yeast in 1921 [1], GSH was found in numerous cellular systems, and is now considered to be a normal and essential constituent of all living cells [2]. GSH is generally the most abundant intracellular nonprotein thiol. For example, in human erythrocytes, GSH is present typically at the 2 to 3 mM level [3], whereas the concentration of ergothioneine, the next most abundant nonprotein thiol in erythrocytes, is estimated to be 0.5 mM [4, 5]. GSSG is also present in the erythrocyte, typically at a concentration some 1% that of GSH [6]. The level of GSH in plasma and other extracellular fluids is too low to be detected by present methods. In contrast, the plasma concentration of cysteine is around 10 μM [7].

1.3. Biological Functions of Glutathione

GSH is involved in numerous cellular processes, as described in detail in the proceedings of two recent conferences devoted to its chemistry and biochemistry [8, 9]. These processes include the protection of cellular membranes and constituents from oxidation by hydroperoxides [10], and the probable detoxification of a variety of electrophilic compounds by the formation of mercapturic acids [11]. An important function of GSH is to keep sulfhydryl groups in their active, reduced state [3]. In this respect, the recent finding [12] that some 80% of the penicillamine in erythrocytes from rheumatoid arthritis patients on penicillamine therapy is in the reduced form, compared to 30% of that in the plasma, is of particular interest in view of the widespread use of penicillamine in chelation therapy. GSH is thought to be involved in the transport of amino acids across cell membranes via the γ-glutamyl cycle [13] and also seems to play a role in determining the behavior of some metal ions in the body. For example, the uptake of $CH_3Hg(II)$ by the rat was found to be dependent on cellular GSH levels [14], suggesting the involvement of the $CH_3Hg(II)$-glutathione complex in the toxicology of $CH_3Hg(II)$. Also, studies with mice have indicated the major biliary metabolite of $CH_3Hg(II)$ to be its GSH complex [15, 16].

2. REDOX CHEMISTRY OF GLUTATHIONE

The sulfhydryl group of GSH is readily oxidized, and its oxidation by O_2 is catalyzed by traces of metal ions such as Cu^{2+}, Fe^{3+}, Co^{2+}, and Mn^{2+} [17]. The $E°'$ for the GSSG-GSH couple

$$GSSG + 2H^+ + 2e^- \rightleftharpoons 2GSH \tag{1}$$

has proven difficult to measure due to the formation of stable thiol-metal complexes at electrode surfaces. For example, in the polarography of GSH, the oxidation wave is due to the oxidation of Hg, with subsequent formation of Hg(I)- and Hg(II)-GSH complexes, rather than the oxidation of GSH itself [18]. Values ranging from -0.35 V to +0.04 V have been reported for $E^{\circ\prime}$ [19]. The value generally considered to be correct, -0.24 V at pH 7.0, was calculated from the equilibrium constant for the reaction

$$GSSG + NADH_2 \rightleftharpoons 2GSH + NAD$$

and the $E^{\circ\prime}$ of the NAD-$NADH_2$ couple [19].

The rate of metal-catalyzed oxidation generally increases with pH, which is interpreted to indicate the involvement of the deprotonated sulfhydryl group [20]. The rate also increases with an increase in the concentration of metal ion. The mechanisms proposed for the metal-catalyzed oxidation all involve the formation of GSH complexes and are compatible with the idea that such complexes are the reactive species. Thus, complexation of Cu^{2+} by EDTA inhibits catalysis [21]. However, some complexes also possess catalytic activity. For example, chelation of Cu^{2+} by o-phenanthroline and by histidine greatly enhances its catalytic effect on the oxidation of GSH [21, 22]. Similarly, copper in caeruloplasmin [23] and cobalt in vitamin B_{12} derivatives [24, 25] are both catalysts for the oxidation of sulfhydryl groups by O_2. Metal ions such as Cu^{2+}, Fe^{3+}, Cr(VI), and Mo(VI) can oxidize sulfhydryl groups directly, also by mechanisms thought to involve inner sphere complex formation [26-29]. GSSG can be reduced directly by some metals, for example, Cu(I) and Hg_2^{2+} [30, 31].

In erythrocytes, glutathione is present in both its reduced and oxidized forms. GSH is oxidized to GSSG as it consumes hydroperoxides, the reaction being catalyzed by glutathione peroxidase. Glutathione reductase in turn catalyzes the conversion of GSSG to GSH, thereby maintaining the GSH-GSSG status of the cell.

3. ACID-BASE CHEMISTRY OF GLUTATHIONE

3.1. Reduced Glutathione

GSH has four groups that undergo acid-base reactions in the pH range 1 to 13. Literature values for the acid dissociation constants are listed in Table 1 [32-38]. By comparison with acid dissociation constants for model compounds, $K_{H_4L}^H$ and $K_{H_3L}^H$ can be assigned to the carboxylic acid groups and $K_{H_2L}^H$ and K_{HL}^H to the other two groups.

TABLE 1

Acid Dissociation Constants of GSH and GSSG

	GSH							GSSG	
$pK_{H_6L}^H$								1.60[h]	
$pK_{H_5L}^H$								2.42	
$pK_{H_4L}^H$	2.12[a]	2.05[b]				2.60[f]	2.04[g]	3.03	3.15[d]
$pK_{H_3L}^H$	3.53	3.40		3.59[d]	3.46[e]	3.82	3.54	4.04	4.03
$pK_{H_2L}^H$	8.66	8.72	8.74[c]	8.75	8.64	9.16	8.54	8.97	8.57
pK_{HL}^H	9.62	9.49	9.62	9.65	9.50	9.88	9.42	9.70	9.54

[a] 22°C [32].

[b] I = 0.2-0.5, 25°C [33].

[c] I = 0.16, 25°C [34].

[d] I = 0.15, 25°C [35].

[e] I = 0.3, 25°C [36].

[f] I = 3.00, 25°C [37].

[g] I = 0.15, 37°C [38].

[h] I = ~0.7, 25°C [45].

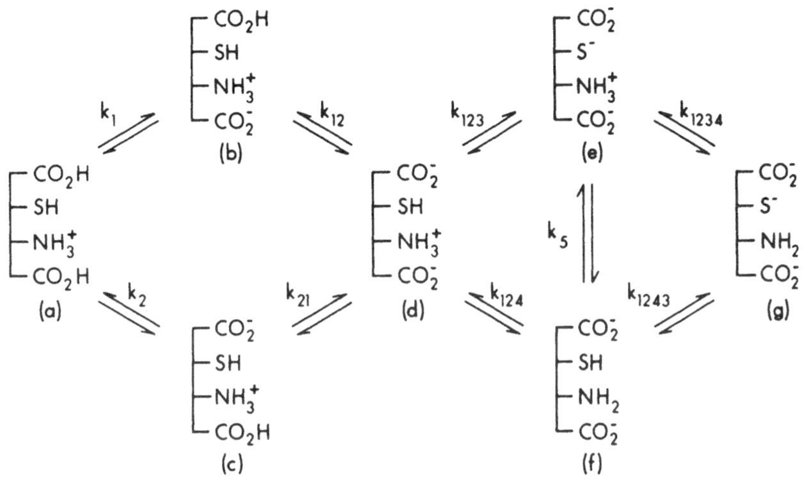

FIG. 1. Microscopic ionization scheme for glutathione. See text for labeling of the acidic groups and definition of the subscripts. (Reprinted with permission from D. L. Rabenstein, *J. Amer. Chem. Soc., 95,* 2797 (1973). Copyright by the American Chemical Society.)

However, the two carboxylic acid groups are of similar acidity, as are the ammonium and sulfhydryl groups, so that these constants are mixed constants and cannot be further assigned to individual groups.

At the molecular level, the acid-base chemistry of GSH is described by eight microscopic constants as shown in the microscopic scheme in Fig. 1 [33, 34]. The equilibrium to which a given constant refers is indicated by the subscript; the last number in the subscript denotes the group involved in the equilibrium under consideration while any preceding numbers denote groups from which protons have already ionized. The carboxylic acid of the L-glutamyl residue is group 1, that of the glycyl residue is group 2, the sulfhydryl group is group 3, and the ammonium group is group 4.

Martin and Edsall [34] obtained values for the four microscopic constants for the sulfhydryl and ammonium groups (Table 2) by pH titration by assuming the ionization constant of the ammonium group of S-methylglutathione to equal k_{124_1} of GSH. All eight microscopic constants have been determined from ^{1}H nmr chemical shift titration

TABLE 2

Microscopic Acid Dissociation Constants of GSH

pk_1		2.19^b
pk_2		3.22
pk_{12}		3.45
pk_{21}		2.42
pk_{123}	8.92^a	8.97
pk_{124}	9.20	9.17
pk_{1234}	9.44	9.35
pk_{1243}	9.16	9.08
k_5	1.9	1.6

^{a}I = 0.16, 25°C [34].
^{b}I = 0.2-0.5, 25°C [33].

curves (Table 2) [33]. The constants obtained by nmr are in good agreement with those reported by Martin and Edsall. From these two sets of constants [33, 34] and the relation $k_5 = k_{123}/k_{124}$, k_5 is calculated to be 1.6 to 1.9, i.e., the sulfhydryl group is 1.6 to 1.9 times as acidic as the ammonium group. The microscopic constants for the carboxylic acid groups indicate the carboxylic acid group of the L-glutamyl residue to be 10.8 times as acidic as that of the glycyl residue. The microscopic constants of the carboxylic acid groups are also reported in a recent ^{1}H nmr study of the conformation of GSH [39].

The microscopic constants for the SH and NH_3^+ groups have also been estimated from spectrophotometric [40, 41] and calorimetric [42] data. It is difficult to assess the validity of the spectrophotometric results because the apparent degree of deprotonation of the GSH sulfhydryl group at a given pH varies with wavelength [34], and

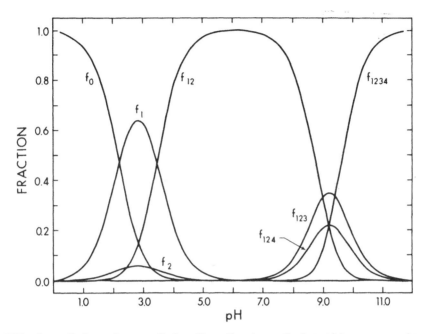

FIG. 2. pH dependence of the distribution of glutathione among its various protonated forms. f refers to the fractional concentration of the species indicated by the subscripts; see text for definition of subscripts. (Reprinted with permission from D. L. Rabenstein, *J. Amer. Chem. Soc.*, *95*, 2797 (1973). Copyright by the American Chemical Society.)

the calorimetric results show considerable variation depending on the data used [42]. In view of the problems associated with the calorimetric method of determining microscopic constants [43, 44], these results are probably not reliable.

The distribution of GSH among its various protonated forms, as calculated from the microscopic constants determined by nmr [33], is shown as a function of pH in Fig. 2. The quantity f refers to the fraction of GSH present at the stage of deprotonation indicated by the subscript, where the subscripts are used in the same manner as with the microscopic constants. The distribution curves in Fig. 2 are of interest with respect to the formation of protonated complexes. As is discussed later, microscopic formation constants can be calculated for protonated complexes if the microscopic acid dissociation

constants, the macroscopic formation constant, and the protonation
site are known.

3.2. Oxidized Glutathione

GSSG has six groups that undergo acid-base reactions in the pH range
1 to 13. The acid dissociation constants of GSSG are listed in
Table 1 [33, 45]. $K_{H_6L}^H$ to $K_{H_3L}^H$ are composite constants for the four
carboxylic acid groups. It has not yet been possible to determine
microscopic constants that can be assigned to the individual groups.
However, chemical shift studies by ^{13}C nmr indicate the carboxylic
acid groups of the two γ-L-glutamyl residues to be the two most
acidic groups [45]. $K_{H_2L}^H$ and K_{HL}^H are composite constants for the
deprotonation of the two ammonium groups. Because of the symmetry
of the H_2L^{2-} species, $k_{12345} = k_{12346}$ and $k_{123456} = k_{123465}$, where
the subscripts 1 to 4 indicate deprotonation of the carboxylic acid
groups and 5 and 6 the ammonium groups. The microscopic constants
can be calculated from the macroscopic constants by the equations
$k_{12345} = K_{H_2L}^H/2 = 5.4 \times 10^{-10}$ and $k_{123456} = 2K_{HL}^H = 4.0 \times 10^{-10}$.

4. COORDINATION CHEMISTRY OF REDUCED GLUTATHIONE

4.1. Zinc

The binding of Zn(II) by GSH has been the subject of a number of
studies [35-38, 46-50], mostly by pH titration techniques. Considering the multiple binding sites in GSH and the lack of information provided at the molecular level by pH titration methods, it is
not surprising that there is a lack of agreement as to the model for
the complexation equilibria. The binding of Zn(II) as revealed by
a ^{13}C nmr study [48] will be described first, and then the pH titration studies will be discussed in light of this information.

The chemical shifts of the three carbons of the cysteinyl
residue are shown as a function of pD in Fig. 3, and the chemical

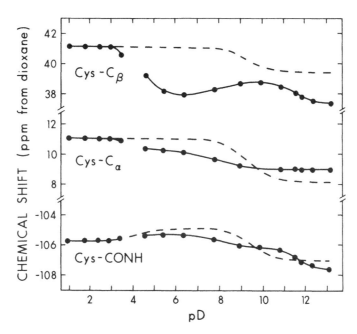

FIG. 3. pD dependence of the chemical shifts of the cysteinyl carbons of GSH in a D_2O solution containing 0.30 M GSH and 0.15 M $Zn(NO_3)_2$. The dashed curves give the chemical shift behavior in a D_2O solution containing no complexing metal ion. (Reprinted with permission from B. J. Fuhr and D. L. Rabenstein, *J. Amer. Chem. Soc.*, *95*, 6944 (1973). Copyright by the American Chemical Society.)

shifts of selected carbon atoms of the glutamyl and glycyl residues are shown for the same conditions in Fig. 4. The dashed curves in Figs. 3 and 4 show the chemical shift vs. pD behavior of the carbon atoms in the absence of complexing metal ion and the solid curves that for a system having a 1:2 Zn(II)-to-GSH ratio.

The basis of the study by nmr of the binding of diamagnetic metal ions by multidentate ligands is that the binding of the metal ion by a particular functional group will cause changes in the chemical shifts of nmr-active nuclei located near the functional group. If the binding is labile on the nmr time scale, as it is in the Zn(II)-GSH complexes, the observed chemical shifts are a weighted average of the chemical shifts of the various complexes.

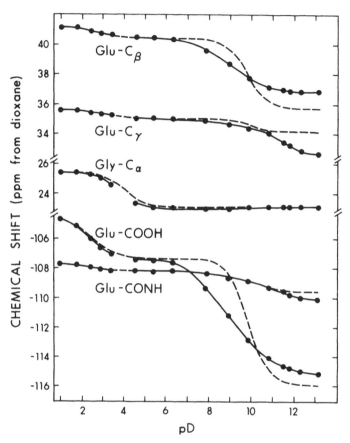

FIG. 4. pD dependence of the chemical shifts of selected carbons of the glutamyl and glycine residues of GSH in a D_2O solution containing 0.30 M GSH and 0.15 M $Zn(NO_3)_2$. The dashed curves give the chemical shift behavior in a D_2O solution containing no complexing metal ion. (Reprinted with permission from B. J. Fuhr and D. L. Rabenstein, *J. Amer. Chem. Soc.*, *95*, 6944 (1973). Copyright by the American Chemical Society.)

At pD < 3, the Zn(II) causes very little change in the ^{13}C chemical shifts. The small differences observed for the $Gly-C_\alpha$, Gly-COOH, $Glu-C_\alpha$, and Glu-COOH chemical shifts indicate a small amount of binding to the carboxylic acid groups, which are shown in Fig. 2 to be at least partially deprotonated by pH > 0. Some binding to the carboxylate groups is consistent with the known binding of

Zn(II) by the carboxylate group of acetylglycine [51] and the
C-terminal end of polyglycine peptides [52].

Between pD 3.0 and 13.2, the chemical shifts of the cysteine
carbon resonances are different in the presence of Zn(II), indicating
that Zn(II) is bound to some extent to the sulfhydryl group over this
pD range. The change in the chemical shift of the Cys-CONH suggests
that there may be some binding to the peptide linkage between the
cysteine and glycine residues, but no conclusions can be drawn
because the perturbations caused by binding of Zn(II) to the sulfur
are expected to be transmitted to carbons several bonds removed.
The chemical shift data for Glu-CONH indicates no binding to the
peptide linkage between the glutamyl and cysteinyl residues at
pD < 10.5. Mole ratio studies at pH 5.51 indicate that the binding
is almost completely to the sulfhydryl group at Zn-GSH mole ratios
$\leq$ 0.5 which, taken with the results in Figs. 3 and 4, leads to the
conclusion that the predominant species in a solution containing
Zn(II) and GSH at a 1:2 ratio at pD 6 has the structural formula:

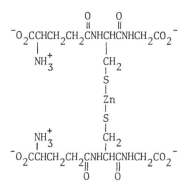

The coordination around the Zn(II) is probably tetrahedral, with
two water molecules occupying the other coordination positions.
That the amino group is still protonated is indicated by the Glu-C_β
and Glu-COOH chemical shift data. These results are in good agreement with the model outlined by Martin and Edsall [47] in an early
study of the complexation chemistry of GSH.

As the pD is increased above 6, Zn(II) causes changes in the
chemical shifts of both the Glu-C_β and the Glu-COOH carbons, indi-

TABLE 3

Formation Constants of M(II)-Glutathione Complexes

Metal	$\log K^M_{M(HL)}$	$\log \beta^M_{M(HL)_2}$	$\log K^M_{ML}$	$-\log K^H_{M(HL)_2}$	$-\log K^H_{M(HL)}$ (L)	$-\log K^H_{ML}$	$-\log K^H_{ML_2}$
Zn(II)[a]	4.74	9.76	7.94	7.04	9.15	8.82	9.86
Zn(II)[b]	4.88	10.86	8.57	7.35	9.68	8.64	9.96
Zn(II)[c]	5.00	10.17		7.15	8.96		
Cd(II)[a]	5.77	11.61	8.59	7.64	9.42	9.12	
Cd(II)[b]	7.14	13.27	10.18	7.94	9.74	9.89	12.18
Cd(II)[c]	6.16	12.06		9.36			
Pb(II)[b]	7.26	12.34	10.57	7.44	9.66		10.50
Pb(II)[c]	6.74	11.93		9.39			

[a] I = 0.15, 37°C. Also $\beta^{Zn}_{Zn_2L}$ = 10.62; $\beta^{Cd}_{Cd_2L}$ = 13.29 [38].

[b] I = 3.00, 25°C [37].

[c] I = 0.30, 25°C. Also $\beta^{Zn}_{Zn(HL)_3}$ = 14.01; $-\log K^H_{Zn(HL)_3}$ = 8.60; $\beta^{Cd}_{Cd(HL)_3}$ = 15.91; $\beta^{Cd}_{Cd_2(HL)_3}$ = 21.70; $\beta^{Pb}_{Pb(HL)_3}$ = 15.63; $-\log K^H_{Pb(HL)_3}$ = 8.63; $\beta^{Pb}_{Pb_2(HL)_3}$ = 21.72 [36].

cating binding to the $^-O_2CCHNH_2$ moiety of the γ-L-glutamyl residue.
From the chemical shift data, it is not possible to determine if
the sulfhydryl and glutamyl groups of a given GSH are simultaneously
coordinated to the same metal ion or if polynuclear complexes form
in which sulfhydryl-complexed Zn(II) ions bind to the glutamyl group
of other sulfhydryl-complexed GSH molecules.

At pD > 10.5, the presence of Zn(II) causes further changes in
the chemical shifts of the Cys-CONH, Cys-C_β, Glu-CONH, and Glu-C_γ
carbons. There are two possible changes in the nature of the complexes that might cause these chemical shift changes: the formation
of hydroxy complexes or the deprotonation of the peptide nitrogen
with subsequent binding to Zn(II). Although the nature of the binding
which causes the chemical shift changes at pD > 10.5 is still an open
question, the data seem most consistent with binding to a deprotonated
peptide linkage [48].

The equilibrium constants obtained for Zn(II)-GSH complexes by
analysis of pH titration data are summarized in Table 3. The equilibrium constants are defined differently than in the original literature
and were calculated from the original values using the acid dissociation constants determined for GSH in the respective studies.

From the results in Table 3, a lack of agreement as to which
Zn(II)-GSH complexes form as well as to the magnitude of their formation constants is apparent. The chemical shift data indicate that,
at pH < 6, the sulfhydryl group is the main binding site while the
amino group is still protonated giving rise to the species Zn(HL)
and $Zn(HL)_2^{2-}$. With the exception of the early study by Li and
coworkers [35] (not listed in Table 3), Zn(HL) and $Zn(HL)_2^{2-}$ have
been included in the models [36-38, 47] and comparison of the formation constants for these complexes indicates reasonable agreement
for $K_{Zn(HL)}^{Zn}$ but somewhat poorer agreement for $\beta_{Zn(HL)_2}^{Zn}$. In the three
studies listed in Table 3, the complexes $Zn(HL)(L)^{3-}$ and $Zn(L)_2^{4-}$ are
also considered to be present. However, the models used by both
Perrin and Watt [38] and Corrie and coworkers [37] differ from that
deduced by Rabenstein and Guevremont [36] in that they also included
a species ZnL^-. Species distribution curves for the Perrin and Watt

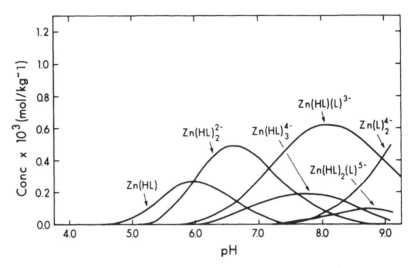

FIG. 5. pH dependence of the distribution of Zn(II) among its various complexed forms in a solution containing 9.78×10^{-4} M $Zn(NO_3)_2$ and 3.86×10^{-3} M GSH. I = 0.30 M $NaClO_4$.

model predict that ZnL^- is a major species in the pH region 6 to 8 in a solution having a Zn(II)-GSH ratio of 1:2, to the extent that greater than 50% of the Zn(II) is in this form at pH 7. The fractional concentration predicted for ZnL^- by the model of Williams et al. is somewhat less, but still more than is consistent with ^{13}C chemical shift data. The models of Perrin and Watt [38] and Williams [37] also include complexes resulting from further deprotonation of ZnL^- and ZnL_2^{4-}. The pK values listed in Table 3 for these two complexes are at least two orders of magnitude too small to correspond to the changes in the ^{13}C spectra attributed to further deprotonation at pD > 10.5.

The distribution of Zn(II) among its various forms as predicted by the constants obtained by Rabenstein and Guevremont [36] is shown as a function of pH in Fig. 5. The first two complexes to form are Zn(HL) and $Zn(HL)_2^{2-}$, in which the Zn(II) is bonded predominantly to the deprotonated sulfhydryl group [48]. Martin and Edsall [47] estimated some 80% of the Zn(II) in Zn(HL) is bonded to the sulfhydryl group with 20% bonded to the $^-O_2CCHNH_2$ group of the γ-L-glutamyl residue by comparison of the binding of Zn(II) by GSH and by S-methylglutathione, where the binding is restricted to the $^-O_2CCHNH_2$ group. At high pH, the predominant species are $Zn(HL)(L)^{3-}$ and then ZnL_2^{4-}, which can be considered to result from titration of a proton from $Zn(HL)_2^{2-}$ and $Zn(HL)(L)^{3-}$ with pK_A values of 7.15 and 8.96, respectively [36]. The pK_A for the titration of a proton from $Zn(HL)_2^{2-}$ is some two orders of magnitude less than the pk for the ammonium group (pk_{124} in Table 2) suggesting some coordination to the $^-O_2CCHNH_2$ group of the γ-L-glutamyl residue. This is as predicted by the ^{13}C nmr results [48].

4.2. Cadmium

The binding of Cd(II) by GSH has also been studied by both the ^{13}C nmr method [48] and pH titration methods [36-38]. The ^{13}C nmr results [48] indicate the complexation of Cd(II) by GSH to be very similar to that of Zn(II). The major complexes at pH < 7 are $Cd(HL)$ and $Cd(HL)_2^{2-}$, in which the Cd(II) is coordinated to the sulfhydryl groups and the amino groups are protonated. The Cd(II) binds more strongly to the sulfhydryl group than does Zn(II). There is also weak binding to the carboxylate groups, and at pH > 7 to the amino group of the glutamyl residue. However, the extent to which the $^-O_2CCHNH_2$ group binds to Cd(II) in solutions having a Cd(II)-GSH ratio of 1:2 is somewhat less than in the case of Zn(II).

Formation constants for the Cd(II)-GSH system are listed in Table 3. In all three studies, $K_{M(HL)}^M$ and $\beta_{M(HL)_2}^M$ are larger than for Zn(II), consistent with the stronger binding to the sulfur indicated in the nmr study. However, the models used by Perrin and Watt [38] and by Williams et al. [37] include the species CdL^-, which is predicted in both studies to be present at concentrations greater than expected from the nmr results. To account for pH titration data, Rabenstein and Guevremont [36] found it necessary to include the polynuclear species $Cd_2(HL)_3$, in which the sulfhydryl and glycyl carboxylate groups of a single GSH might be bonded simultaneously to different metal ions [46]. Similar polynuclear complexes have been reported for the Ag(I)-GSH [53, 54] and Hg-GSH [55, 56] systems.

The Cd(II)-GSH system has also been studied by ^{113}Cd nmr [57-59]. Although these studies do not provide any new insight to the nature of the Cd(II)-GSH complexes, they suggest that ^{113}Cd nmr will be of use in identifying the groups bonded to Cd(II) in Cd(II)-protein complexes.

Williams and coworkers [37] concluded from calculations on the equilibrium concentrations of Cd(II)-GSH and Zn(II)-GSH complexes under conditions intended to simulate those of plasma that GSH might be a good drug for increasing the rate of elimination of Cd(II) from the body. However, as mentioned in the introduction, the level of GSH in the plasma is too low to detect with current methods, and it is to be expected that GSH administered in chemotherapy would be consumed in normal metabolic processes. Thus, GSH has been found to be ineffective in increasing the rate of elimination of $CH_3Hg(II)$ from mice injected with CH_3HgCl [60], and has been found to increase Cd concentrations in the kidney and testis of mice [61].

4.3. Lead

^{13}C nmr results indicate that, in solutions having a Pb(II)-GSH ratio of 1:2, the binding of Pb(II) by GSH is considerably more selective than is the binding of Zn(II) or Cd(II), with Pb(II) binding only to the sulfhydryl group and, to a lesser extent, to the glycine carboxylate group [48]. The extent of binding to the carboxylate group decreases at pD > 9, presumably due to displacement by hydroxide ions with the formation of Pb(II)-hydroxy-GSH mixed complexes. The chemical shift data for the Glu-COOH and Glu-C_α carbons indicates no detectable binding to the amino group, with the acid-base properties of the amino group in Pb(HL) and Pb(HL)$_2^{2-}$ virtually identical to the acid-base properties of the amino group of free GSH.

The formation constants listed in Table 3 [36, 37] for the Pb(HL) and Pb(HL)$_2^{2-}$ complexes indicate the binding to the sulfhydryl group in both complexes to be similar to that in the analogous Cd(II) complexes and slightly stronger than in the Zn(II) complexes. Williams and coworkers [37] also consider the complex PbL$^-$ to be present, and their constants predict it to be a major species in the pH range 6 to 9 for a solution having a Pb-GSH ratio of 1:2. There is no evidence for this species in the ^{13}C nmr results. The pK$_A$ value of 9.39 [36] for Pb(HL)$_2^{2-}$ is in accord with the lack of any effect on the acidity of the ammonium group as indicated by the nmr study [48]. It has been concluded from enthalpy and entropy data [62] that the amino group is coordinated in the Pb(II)-GSH complexes, but this conclusion must be taken with some reservation because it was concluded in the same study that the amino group is complexed by Pb(II) in the monoprotonated complexes of glycine, diglycine, and triglycine. ^{1}H nmr results show the structural formulas to be Pb(II)-$^-$O$_2$C(CH$_2$NHCO)$_x$CH$_2$NH$_3^+$ [51, 52].

Assuming the conclusion that Pb(II) binds selectively to the sulfhydryl group in the complexes Pb(HL) and Pb(HL)$_2^{2-}$ is correct, microscopic formation constants can be calculated from the constants in Table 3 and the microscopic acid dissociation constants for GSH. The microscopic formation constants, $k_{Pb(HL)}^{Pb}$ and $k_{Pb(HL)_2}^{Pb}$, are defined in terms of Pb(II) in equilibrium with the monoprotonated isomer e in Fig. 1 and are related to the constants in Table 3 by the relations $k_{Pb(HL)}^{Pb} = K_{Pb(HL)}^{Pb} \times K_{H_2L}^{H}/k_{123} = 1.2 \times 10^7$ and $k_{Pb(HL)_2}^{Pb} = \beta_{Pb(HL)_2}^{Pb} \times (K_{H_2L}^{H}/k_{123})^2 = 3.9 \times 10^{12}$. The advantage of constants defined in this way is that they more correctly reflect the stability of the complex.

It is of interest that the red cell levels of GSH are somewhat reduced in lead-exposed workers [63].

The binding of trimethyllead, (CH$_3$)$_3$Pb(IV), by GSH has also been studied by ^{1}H nmr [64]. (CH$_3$)$_3$Pb(IV) tends to bind to a single donor group [65], and the ^{1}H nmr results indicate weak binding to the carboxylate groups, little or no binding to the amino group, and the strongest binding to the deprotonated sulfhydryl group. The binding is, however, somewhat weaker than in analogous complexes of Pb(II).

4.4. Mercury

As would be expected from the strong affinity of sulfur for mercury
and the tendency of Hg(II) toward linear two coordination, Hg(II)
binds to two GSH ligands through their deprotonated sulfhydryl groups
over the pH range 1 to 11 in solutions having a GSH-Hg(II) ratio of
2 or more [48, 55]. As the pH is changed, the carboxylic acid and
ammonium groups titrate with essentially the same acid-base properties as in the absence of Hg(II) [48], indicating no participation
of these groups in binding to Hg(II) in the 2:1 complex. From emf
measurements of Hg metal in equilibrium with solutions containing
Hg(II) and a large excess of GSH, Stricks and Kolthoff [55] deter-
mined the formation constants $\log \beta_{Hg(HL)_2}^{Hg} = 41.0$, $\log \beta_{Hg(HL)(L)}^{Hg} =$
41.9, and $\log \beta_{HgL_2}^{Hg} = 41.6$.

In the absence of chloride ion, GSH and Hg(II) can also form
the complexes $Hg_2(GS)_2$ and $Hg_3(GS)_2$, where GS represents sulfhydryl-
coordinated GSH, over the pH range 3 to 9 [55, 56]. One Hg(II) in
each complex is bound firmly as the mercaptide (GS-Hg-SG) while the
others are more loosely bound, probably by way of the amino and
carboxylate groups. In the presence of excess chloride, the loosely
bound Hg(II) is complexed as $HgCl_4^{2-}$. It has been proposed on the
basis of ^{13}C chemical shifts that Hg(II) forms a chelate with GSH
by binding to the sulfhydryl group and a deprotonated peptide nitrogen in Hg(II)-GSH complexes of 1:1 stoichiometry [66]. However, it
seems more reasonable to interpret the small shifts attributed to
binding to the peptide nitrogen in terms of changes in the extent of
protonation of the carboxylic acid groups due to lack of pH control
and to through bond effects resulting from binding to the sulfur.

Mercurous acetate reduces the disulfide bond of GSSG according
to the scheme [31].

$$GSSG + Hg_2^{2+} \rightleftharpoons \left\{ \begin{array}{c} 2(GSHg(II)) \\ \uparrow\downarrow \\ GSHgSG + Hg^{2+} \end{array} \right\}$$

The time scale of this reaction is such that 65% of the disulfide
groups are reduced after 4 days. From the standard electrode potentials for the GSSG-GSH and Hg^{2+}-Hg_2^{2+} couples, reduction of the
disulfide bond by Hg_2^{2+} is predicted to be an unfavorable process,
but the standard free energy associated with the formation of the
Hg(II)-sulfhydryl bond offsets that for the redox step and the
reduction of GSSG by Hg_2^{2+} does indeed take place.

   Also because of the strong binding of Hg by the sulfhydryl
group, the anodic wave in the polarography of GSH (E 1/2 of -0.38 V
vs. SCE at pH 7) is due to depolarization of the Hg electrode with
the formation of Hg(I)-GSH complex rather than the oxidation of the
sulfhydryl group (GSH + Hg $\rightleftarrows$ GSHg + H^+ + e^-) [18]. The mercurous-GSH
complex is unstable and readily decomposes to Hg metal and Hg(II)-GSH
complexes. Very few compounds in biological fluids other than those
containing sulfhydryl groups give anodic currents at such low potentials, and this selectivity has been made the basis of a detection
system for use in liquid chromatographic methods for the determination
of GSH and other sulfhydryl compounds in blood [12, 67].

   The binding of $CH_3Hg(II)$ by GSH has been studied in some detail
as a model for its binding by biological molecules [33, 68-70].
$CH_3Hg(II)$ generally complexes as a one coordinate species to give
linear, two-coordinate Hg(II) [70]. In the 1:1 $CH_3Hg(II)$-GSH complex,
$CH_3Hg(II)$ binds to the deprotonated sulfhydryl group with no detectable dissociation over the pH range 0 to 14 [33, 68]. The microscopic
acid dissociation constants of the two carboxylic acid groups and the
ammonium group of the 1:1 complex are essentially identical to those
of free GSH [33], indicating no participation of these groups in the
binding of $CH_3Hg(II)$. Simpson has estimated the formation constant
of the complex to be 7.9 x 10^{15} [71]. The formation constants reported for sulfhydryl complexes in a recent study [72] of $CH_3Hg(II)$
are considerably smaller than this value, and much too small to be
consistent with the strong binding demonstrated in the nmr studies
[33, 68]. $CH_3Hg(II)$ forms higher complexes with GSH, and the nature
of the complex is strongly dependent on pH [68]. For example, in

the 2:1 $CH_3Hg(II)$-GSH complex, both $CH_3Hg(II)$ ions are bonded to the sulfhydryl group up to pH 4, with the amino group protonated. As the pH is increased from 4 to 8, one $CH_3Hg(II)$ shifts to the ammonium group, displacing a proton, where it binds up to pH 10. At pH > 10, the loosely bound $CH_3Hg(II)$ dissociates with the formation of CH_3HgOH.

The kinetics of the ligand exchange reactions of CH_3HgSG have been characterized in the pH range 0.5 to 3 by nmr line-broadening techniques [68]. The predominant exchange pathways are

$$CH_3HgSG + H^+ \underset{k_{-1}}{\overset{k_1}{\rightleftharpoons}} CH_3Hg(II) + GSH \tag{2}$$

and

$$CH_3HgSG + \overset{*}{G}S^- \overset{k_2}{\rightleftharpoons} CH_3Hg\overset{*}{S}G + GS^- \tag{3}$$

The rate constants are k_1 = 600 ± 200, k_{-1} = 5.1 x 10^9, and k_2 = (5.8 ± 1.9) x 10^8 liter/mol sec. The apparent mobility of $CH_3Hg(II)$ in biological systems, which perhaps is surprising in view of the large stability constants for its binding by sulfhydryl groups, can be accounted for with this system as a model [69]. At pH 7, the ligand exchange process represented by Eq. (3) is most important and predicts that $CH_3Hg(II)$ will exchange among the various SH-containing ligands many times per second at SH concentrations in the range of those in plasma and red cells [69]. The extent to which $CH_3Hg(II)$ is taken up by the red cells of rats depends on the intracellular GSH level [14], suggesting that the CH_3HgSG complex does indeed play a major role in determining the behavior of $CH_3Hg(II)$ in biological systems.

GSH complexes of a variety of organomercurials have been studied as models for their reactions with the sulfhydryl groups of proteins [73-78].

4.5. Silver

Ag(I) reacts with GSH to form complexes of variable stoichiometry,
depending on the solution conditions. Polarographic and potentiometric results indicate the existence of compounds of 1:1 and 1:2
(Ag-GSH) stoichiometry [79, 80]. In tris buffer at pH 7 to 8, a
complex of 1:2 stoichiometry forms in which the Ag(I) is bonded to
the two deprotonated sulfhydryl groups (GS-Ag-SG) whereas the amino
groups are presumably largely protonated [53]. When the Ag-GSH
ratio is increased to 1:1, a complex thought to be a polymer of
1:1 stoichiometry forms. The degree of polymerization is dependent
on the concentration of Ag(I) in the solution during the polymerization up to an Ag(I)-GSH ratio of 1:1. Sedimentation studies on a
pH 8 solution containing 0.010 M GSH, 0.008 M $AgNO_3$, and 0.10 M tris
buffer indicate a molecular weight of 6600 ± 700. The binding in
the polymeric complex is thought to involve the amino groups as well
as the sulfhydryl groups. Equilibrium constants for protonated and
polynuclear Ag(I)-GSH complexes have also been reported [54].

    Considerable use has been made of the reaction of Ag(I) with
the sulfhydryl group of GSH in methods for detecting and determining
GSH in blood and tissues [81, 82]. An ammoniacal titration medium
is used to control the stoichiometry of the reaction.

4.6. Nickel

In contrast to the selective binding of heavy metals by the sulfhydryl
group of GSH, this group is a minor binding site for Ni(II) for most
solution conditions. Martin and Edsall [47] showed by comparison of
the formation constants of the Ni(II) complexes of GSH and S-methyl
GSH that in the GSH complexes of stoichiometry Ni(HL), the Ni(II) is
predominantly bonded to the γ-glutamyl center. Circular dichroism
(CD) measurements on solutions containing sufficient base to convert

GSH to its HL^{2-} form [83] and ^{1}H nmr studies based on selective line
broadening by paramagnetic Ni(II) [84] also are consistent with a
majority of the binding at the γ-glutamyl center. Analysis of the
pH dependence of the rate of complexation of Ni(II) by GSH in the
pH range 6 to 7 indicates initial complexation to be at the carboxylate group of the $^-O_2CCHNH_3^+$ zwitterion, followed by proton loss from
the ammonium group and chelate ring closure [85].

In the intermediate pH range, the ^{1}H nmr study suggests some
binding to the glycine residue, which was interpreted in terms of a
binding through the glycine carboxylate group and its peptide nitrogen in the deprotonated form [84]. This seems somewhat surprising
in view of the higher pH usually required for Ni(II)-promoted peptide
nitrogen deprotonation [86]. At higher pH, both the CD [83] and [1]H
nmr [84] results indicate a change in the nature of the complexes,
with increased binding to the sulfhydryl group and a change in the
geometry from paramagnetic octahedral to diamagnetic square planar.
A diamagnetic Ni(II) complex of 1:1 stoichiometry has been isolated
[87].

4.7. Copper

Characterization of the complexation of Cu(II) by GSH is complicated
by the ease with which it catalyzes oxidation of the sulfhydryl
groups [26, 88], as described for sulfhydryl groups in general [89]
by:

$$2Cu^{2+} + 2RS^- \rightleftharpoons 2Cu^+ + RSSR$$

The most detailed study of Cu(II) binding has made use of selective
broadening in the ^{1}H nmr spectrum of GSH and electron paramagnetic
resonance (EPR) measurements on Cu(II) [90]. In acidic solution,
Cu(II) causes little broadening of GSH resonances, indicating little
if any binding under these conditions. The CH_2 resonance of the
glycine residue is not affected by Cu(II) over the whole pH range.
At pH > 8, the addition of Cu(II) to GSH solutions first causes

broadening of the ^{1}H resonances from the cysteine residue and, at
higher Cu(II) concentrations, broadening of the γ-glutamyl α-CH
resonance. These results were interpreted to indicate binding first
to the sulfhydryl and deprotonated peptide nitrogen of the cysteinyl
residue, and then to the γ-glutamyl groups. EPR results indicate
nonequivalent Cu(II) ions coordinated to GSH at Cu(II)-GSH ratios
> 1:1, which have been interpreted as one Cu(II) coordinated at the
cysteinyl residue and the other at the γ-glutamyl groups. The correct assignment of binding sites from nmr line-broadening data requires that scalar interactions be accounted for and that ligand
exchange be on the proper time scale [91, 92]. It is not apparent
as to whether the authors [90] have determined this to be the case.
This is of particular interest with respect to the proposed binding
involving the deprotonated peptide nitrogen of the cysteinyl residue.

The binding of Cu(I) by GSH has been reported in a study of
Cu(I) complexes of sulfur-containing ligands [93]. The RS$^-$ group
forms the strongest complexes with Cu(I) [89], and is apparently
the only site to which Cu(I) binds in the Cu(I)-GSH complexes, which
are insoluble [93].

4.8. Cobalt

Evidence suggests that the predominant binding site in the Co(II)
complex of stoichiometry Co(HL) is the $^-O_2CCHNH_2$ group of the
γ-glutamyl residue [47].

Complexation of GSH by Co(III) in the form of cobalamins,
corrinoids, and related model compounds has been the subject of
considerable study [25, 94-101]. In the cobalamins and corrinoids,
four coordination positions of the Co(III) are occupied by the
extremely stable, more or less planar corrin ring. Since the
remaining two coordination sites on Co(III) lie on opposite sides
of the corrin ring, chelation to Co(III) is not possible. Bridged,
binuclear complexes involving Co(III) coordination at the amino and

sulfhydryl groups have been found in studies with model compounds
[99]. Under anaerobic conditions, both the aquo and hydroxo forms
of cobalamin react with GSH to form a single complex of 1:1 stoichiometry [25, 94, 97]. Cyanide readily displaces GSH from the complex
to give dicyanocobalamin [96, 97], and one equivalent of p-mercuri-
benzoate per mole of GSH also prevents complex formation [94]. The
formation constant for the GSH-cobalamin complex in which GSH is
bound through the sulfhydryl group has been estimated to be 7.7×10^4
at pH 7 [100]. Cobalt(II) forms of cobalamins and cobanamides also
form GSH complexes, presumably by coordination to the sulfhydryl
group [102].

GSH has been reported to labilize the Co-C bond of methylcobalamin by promoting homolytic cleavage to give cob(II)alamin
(B^{12}-R and the thioether, and also to coordinate to the Co(III) in
5-deoxyadenosylcobalamin at pH 7.4 and methylcobalamin at pH 4.0 by
displacement of the 5,6-dimethylbenzimidazole group from the sixth
coordination site [102], although a more recent report concludes
that no complex is formed between these alkyl corrinoids and GSH
[101]. Rather, GSH displaces benzimidazole by protonation rather
than competitive coordination.

In the presence of oxygen, aquocobalamin and cyanoaquocobalamin
catalyze the oxidation of GSH to the disulfide [25]. The rate increases with pH, indicating the GS$^-$ anion to be the oxidizable form.
Also, cob(II)alamin is obtained from pH 7.4 solutions of the GSH-
aquocobalamin complex at elevated temperatures [100]. Oxidation of
the sulfhydryl group is indicated by epr signals for sulfhydryl
radicals. By contrast, the Co(III) in diaquocobinamide oxidizes
GSH only when present in considerable excess [100].

4.9. Molybdenum

GSH is oxidized by Mo(VI), with the formation of Mo(V) and GSSG [29].
The kinetics of this reaction have been studied at pH 7.50 in phosphate buffer, and mechanistic details have been proposed. No evidence was obtained for complex formation between Mo(VI) and GSH.

Mo(V), however, forms complexes with GSH which are of interest as models for the Mo(V) in molybdoflavoenzymes such as xanthine oxidase [103, 104]. A dinuclear complex of stoichiometry $[Mo(V)]_2$-GSH in which the molybdenum coordinates as the diamagnetic dioxobridged unit

has been isolated and characterized [104]. In phosphate buffer at pH 8 to 10, this complex dissociates to give an EPR-active, paramagnetic, dinuclear complex.

$$[MO(V)]_2L \text{ (diamagnetic)} \xrightleftharpoons{OH^-} [Mo(V)]_2L \text{ (paramagnetic)}$$

Dissociation is thought to occur in the Mo_2O_4 unit. The solution EPR spectrum of the complex prepared from isotopically enriched ^{95}Mo shows 11 hyperfine lines, as expected for an electron interacting with two magnetically equivalent Mo(V) nuclei of spin 5/2. In the structures proposed for both the diamagnetic [104] and paramagnetic complexes [104, 105], it is somewhat surprising to see binding to protonated peptide nitrogens because they usually show little tendency toward metal coordination [26, 106]. It seems likely they will be revised as the complexes are studied further. These complexes are of considerable interest as models for biological molybdenum in that they meet the requirements of having two molybdenum ions coordinated by a sulfhydryl ligand and of giving rise to a paramagnetic species, as has been observed for xanthine oxidase by EPR measurements [104].

A Mo(V)-GSH complex having a Mo-to-N ratio of 4:3, corresponding to four Mo atoms per GSH ligand, has also been isolated [107]. Although no structural details are available, this complex is of interest because it shows some catalytic activity in the conversion of acetylene to ethylene and hydrazine to ammonia by borohydride.

4.10. Manganese

From the formation constant of the S-methyl GSH complex of Mn(II)
and the estimated microscopic formation constant of the Mn(HL) complex in which Mn(II) is coordinated to the sulfhydryl group of GSH
and the amino group protonated (k = 80), Martin and Edsall [47] concluded that the $^-O_2CCHNH_2$ group of the γ-glutamyl residue is the
major binding site for Mn(II) in the Mn(HL) species. A study of
the Mn(II)-GSH system by EPR has provided evidence of interaction
between Mn(II) and GSH, but it was not possible to separate the
effects seen in the EPR measurements into contributions from outer
sphere and inner sphere complexation [108].

4.11. Iron

Complexation of Fe(III) by GSH in the pH range 1 to 3 has been
studied by kinetic and spectrophotometric methods [109]. For these
conditions, the binding constants for complexes of 1:1 stoichiometry
are described by the equation pK = -1.96 - 0.50 pH, where pK is -log
(formation constant). The binding sites and degree of protonation
of the complexes could not be determined. It is perhaps curious
that apparently no oxidation of GSH was observed.

In a study of models for the iron in ferredoxin, it was found
that ferredoxin-like spectra were obtained by addition of sulfide
to solutions containing Fe(III) and GSH at pH 8.6 [110]. A highly
distorted binuclear structure with octahedral coordination around
each iron and inorganic sulfur bridges was proposed for the GSH-
Fe(III)-sulfide complex. It was also observed that the spectra of
solutions containing only Fe(III) and GSH changed with time, becoming
identical with that of the ferredoxin-like spectrum after some 300
min. This was interpreted to indicate release of sulfur from GSH
in the Fe(III)-GSH complex, which is then incorporated to give the
ferredoxin-like complex [110].

It is of related interest that a considerable fraction of the GSH in red blood cells is apparently associated with oxyhemoglobin [111]. Conversion to the deoxy form releases the GSH. The nature of the interaction is not yet known.

4.12. Gold

The complexation of gold by GSH has been studied in relation to its use in the treatment of rheumatoid arthritis [112-114]. It has been demonstrated by ^{13}C nmr that GSH readily displaces thiomalate from Au(I)-thiomalate and thioglucose from Au(I)-thioglucose [113, 114], the two most widely used gold drugs for rheumatoid arthritis. The GSH exchange reactions on Au(I) are facile, and probably involve a polynuclear mixed ligand cluster of the type Au_4S_6 with GSH binding through its sulfhydryl group [113, 114]. These studies are of considerable interest with respect to elucidating the molecular basis of the pharmacological activity of Au(I)-thiomalate and Au(I)-thioglucose.

4.13. Chromium

Cr(VI) oxidizes GSH to the disulfide and a detailed kinetic study [115, 116] indicates the overall reaction to be

$$nH^+ + 2[HCrO_4]^- + 6GSH \rightleftharpoons 2Cr(III) + 3GSSG \qquad (4)$$

The reaction proceeds via a transient intermediate which can decompose either by a proton-catalyzed pathway or by reaction with a second mole of GSH. The transient intermediate has been identified as a 1:1 GSH-chromate ester [116], for which the formation constant, as defined by Eq. (5), is 1,440 liter/mol at 25°C.

$$GSH + [(HO)CrO_3]^- \rightleftharpoons [(RS)CrO_3]^- + H_2O \qquad (5)$$

GSH binds to the Cr(VI) center via the sulfhydryl group.

4.14. Group IIA Metals

There are few reports in the literature of binding of the alkaline
earth metals by GSH. Touche and Williams [117] reported formation
constants for the calcium-GSH system and for mixed calcium-zinc-GSH
complexes. Martin and Edsall [47] reported that magnesium is bound
only weakly by GSH.

4.15. Lanthanides

In the only reported study of complexation of lanthanides by GSH,
formation constants were determined from pH titration data for
La(III) complexes of GSH and for mixed La(III)-Zn(II)-GSH complexes
[117].

5. COORDINATION CHEMISTRY OF OXIDIZED GLUTATHIONE

The predominant isomer of GSSG at neutral pH is

$$
\begin{array}{c}
 O O \\
 \| \| \\
{}^-O_2CCHCH_2CH_2CNHCHCNHCH_2CO_2{}^- \\
 | | \\
 NH_3^+ CH_2 \\
 | \\
 S \\
 | \\
 S \\
 | \\
 NH_3^+ CH_2 \\
 | | \\
{}^-O_2CCHCH_2CH_2CNHCHCNHCH_2CO_2{}^- \\
 \| \| \\
 O O
\end{array}
$$

Although GSSG presents even more coordination sites than GSH, it
lacks the comparatively reactive sulfhydryl group which plays a
considerable role in the coordination chemistry of GSH and there
are but few reports on the complexation chemistry of GSSG. The
disulfide group does not bind strongly to metal ions, so that the

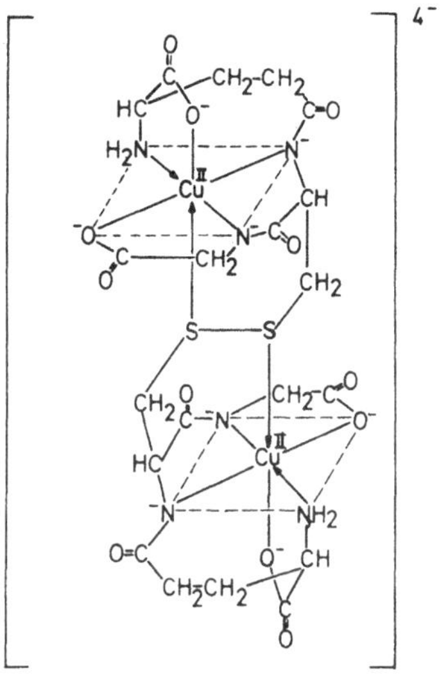

FIG. 6. Proposed structure for the $[Cu(II)]_2$ GSSG complex.
(Reprinted by permission from P. Kroneck, *J. Amer. Chem. Soc.*,
97, 3839 (1975). Copyright by the American Chemical Society.)

most important binding sites in GSSG are the carboxylate and amino
groups.

The Cu(II) complexes of GSSG have been studied as a model for
copper in the "blue oxidases." From polarographic measurements,
Li et al. [118] concluded that GSSG forms only a 1:1 complex with
a formation constant ($[CuL^{2-}]/[Cu^{2+}][L^{4-}]$) of 4×10^{14}. Kroneck
[119] has reported, however, that titration of a 2:1 solution of
Cu(II) and GSSG with base to pH 9.5 is accompanied by the liberation
of eight proton equivalents per mole of GSSG and the formation of a
complex with a single absorption maximum at 590 nm. By the method
of continuous variations, the concentration of the species absorbing
at 590 nm reaches a maximum at a Cu-GSSG ratio of 2:1. The complex
is stable toward hydrolysis up to pH 13, where it decomposes due to
alkaline cleavage of the disulfide bond. With S-methyl GSH, a
stable 1:1 complex is formed. The structure proposed for the 2:1
Cu(II)-GSSG complex is shown in Fig. 6. Each Cu(II) is six-
coordinate, with Cu(II)-Cu(II) interaction via a disulfide bridge,

as necessary to account for the weak EPR signal characteristic of a spin coupled Cu(II) dimer [119]. Another particularly interesting feature of the proposed structure is the Cu(II) binding to all four peptide groups via deprotonated peptide nitrogens. Hopefully the structure of this complex will be determined by X-ray crystallography.

In a study of the loss of Cu metal from intrauterine devices, it was found that GSSG increases the rate of dissolution of Cu [30]. A mechanism proposed for the dissolution in saline solutions containing oxygen involves oxidation of metallic Cu at its surface to give Cu_2O, which then dissolves as a cuprous chloride complex [120]. The Cu(I) then cleaves the disulfide bond, presumably by reduction because SH groups were detected in the reaction mixture.

Silver ion also causes cleavage of the disulfide bond of GSSG according to the reaction [121]:

$$2GSSG + 3Ag^+ + 2H_2O \rightleftarrows 3GSAg + GSO_2H + 3H^+$$

A detailed kinetic study suggests that the reaction proceeds by a direct attack of Ag^+ ion on the disulfide.

Li et al. [35] estimated the formation constant of a Zn(II)-GSSG complex of 1:1 stoichiometry to be 1.7×10^7.

6. SUMMARY

In this chapter we have attempted to review the literature on the binding of metal ions by reduced and oxidized glutathione. Because of the abundance of coordination sites, their coordination chemistry is complicated and the most definitive information has been obtained by using results from a combination of techniques. It would seem that crystal structures of GSH complexes, of which there are none at the present time, would also be useful for identifying the possible binding modes. The complexes formed by GSH with some metal ions have been studied in detail, but little is known about the complexes formed with most metal ions. Even less is known about the dynamics of the complexation reactions.

Considering the abundance of GSH in nature and the affinity of its sulfhydryl group for metals, particularly the heavy metals, it seems likely that GSH complexes form in vivo. To date, the strongest evidence for such complexes comes from studies of the toxicology of methyl mercury. It recently has been demonstrated that the biochemistry of GSH in human erthrocytes can be studied directly at the natural abundance level by ^{1}H nmr spectroscopy by using multiple pulsed Fourier transform techniques [122]. Perhaps with these techniques, the chemistry of such complexes can ultimately be studied in cellular systems.

ACKNOWLEDGMENT

Our research on the chemistry of glutathione has been supported by the National Research Council of Canada and the University of Alberta. It is a pleasure to acknowledge the contributions of our collaborators, whose names appear in the references. We thank Professors R. B. Martin, H. Freeman, and A. McAuley for reading and commenting on the manuscript and the American Chemical Society for permission to use Figs. 1-4 and 6.

ABBREVIATIONS AND DEFINITIONS

The following examples illustrate the definition of the constants given in Tables 1 and 3. For convenience, charges are omitted.

$$H_6L \rightleftharpoons H + H_5L \qquad K^H_{H_6L} = \frac{[H][H_5L]}{[H_6L]}$$

$$HL \rightleftharpoons H + L \qquad K^H_{HL} = \frac{[H][L]}{[HL]}$$

$$M + HL \rightleftharpoons M(HL) \qquad K^M_{M(HL)} = \frac{[M(HL)]}{[M][HL]}$$

$$M + 2(HL) \rightleftharpoons M(HL)_2 \qquad \beta^M_{M(HL)_2} = \frac{[M(HL)_2]}{[M][HL]^2}$$

$$M(HL)_2 \rightleftharpoons M(HL)(L) + H \qquad K^H_{M(HL)_2} = \frac{[M(HL)(L)][H]}{[M(HL)_2]}$$

$$ML \rightleftharpoons M(L - 1H) + H \qquad K^H_{ML} = \frac{[M(L - 1H)][H]}{[ML]}$$

REFERENCES

1. F. G. Hopkins, *Biochem. J.*, *15*, 286 (1921).

2. E. M. Kosower, in Glutathione: Metabolism and Function (I. M. Arias and W. B. Jakoby, eds.), Raven Press, New York, 1976, p. 1.

3. H. K. Prins and J. A. Loos, in Biochemical Methods in Red Cell Genetics (J. J. Yunis, ed.), Academic Press, New York, 1969, p. 115.

4. G. Hunter, *Can. J. Res.*, Section E, *27*, 230 (1949).

5. R. P. McMenamy, C. C. Lund, G. J. Neville, and D. F. H. Wallach, *J. Clin. Invest.*, *39*, 1675 (1960).

6. P. C. Jocelyn, *Biochem. J.*, *27*, 363 (1960).

7. R. Saetre and D. L. Rabenstein, *Anal. Biochem.*, *90*, 684 (1978).

8. L. Flohé, H. Ch. Benöhr, H. Sies, H. D. Waller, and A. Wendel, eds., Glutathione, Academic Press, New York, 1974.

9. I. M. Arias and W. B. Jakoby, eds., Glutathione: Metabolism and Function, Raven Press, New York, 1976.

10. L. Flohé, W. A. Gunzler, and R. Ladenstein in Ref. 9, p. 115.

11. L. F. Chasseaud in Ref. 9., p. 77.

12. R. Saetre and D. L. Rabenstein, *Anal. Chem.*, *50*, 276 (1978).

13. A. Meister in Ref. 9., p. 35.

14. R. Richardson and S. Murphy, *Toxicol. Appl. Pharmacol.*, *31*, 505 (1975).

15. T. Norseth, *Acta Pharmacol. Toxicol.*, *29*, 375 (1971).

16. T. Refsvik and T. Norseth, *Acta Pharmacol. Toxicol.*, *36*, 67 (1975).

17. C. C. Tsen and A. L. Tappel, *J. Biol. Chem.*, *233*, 1230 (1958).

18. W. Stricks and I. M. Kolthoff, *J. Amer. Chem. Soc.*, *74*, 4646 (1952).

19. N. Rost and S. Rapoport, *Nature*, *201*, 185 (1964).

20. P. C. Jocelyn, Biochemistry of the SH Group, Academic Press, New York, 1972, p. 94.

21. K. Kobashi, *Biochem. Biophys. Acta, 158,* 239 (1968).

22. I. G. Fels, *Exp. Eye Res., 12,* 227 (1971).

23. P. C. Jocelyn, unpublished results quoted in Ref. 9, p. 95.

24. J. L. Peel, *J. Biol. Chem., 237,* PC263 (1962).

25. J. Aronovitch and N. Grossowicz, *Biochem. Biophys. Res. Comm., 8,* 416 (1962).

26. R. B. Martin, in Metal Ions in Biological Systems (H. Sigel, ed.), Marcel Dekker, New York, Vol. 1, 1974, p. 129.

27. K. J. Ellis, A. G. Lappin, and A. McAuley, *J. Chem. Soc. Dalton,* 1930 (1975).

28. A. McAuley and M. A. Olatunji, *Can. J. Chem., 55,* 3335 (1977).

29. J. F. Martin and J. T. Spence, *J. Phys. Chem., 74,* 2863 (1970).

30. G. K. Oster, *Nature, 234,* 153 (1971).

31. M. M. David, R. Sperling, and I. Z. Steinberg, *Biochim. Biophys. Acta, 359,* 101 (1974).

32. N. W. Pirie and K. G. Pinhey, *J. Biol. Chem., 84,* 321 (1929).

33. D. L. Rabenstein, *J. Amer. Chem. Soc., 95,* 2797 (1973).

34. R. B. Martin and J. T. Edsall, *Bull. Soc. Chim. Biol., 40,* 1763 (1958).

35. N. C. Li, O. Gawron, and G. Bascuas, *J. Amer. Chem. Soc., 76,* 225 (1954).

36. D. L. Rabenstein and R. Guevremont, submitted for publication.

37. A. M. Corrie, M. D. Walker, and D. R. Williams, *J. Chem. Soc. Dalton,* 1012 (1976).

38. D. D. Perrin and A. E. Watt, *Biochim. Biophys. Acta, 230,* 96 (1971).

39. S. Fujiwara, G. Formicka-Kozlowska, and H. Kozlowski, *Bull. Chem. Soc. Jap., 50,* 3131 (1977).

40. R. E. Benesch and R. Benesch, *J. Amer. Chem. Soc., 77,* 5877 (1955).

41. D. M. E. Reuben and T. C. Bruice, *J. Amer. Chem. Soc., 98,* 114 (1976).

42. D. L. Vander Jagt, L. D. Hansen, E. A. Lewis, and L. B. Han, *Arch. Biochem. Biophys., 153,* 55 (1972).

43. E. Coates, C. G. Marsden, and B. Rigg, *Trans. Faraday Soc., 65,* 3032 (1969).

44. E. W. Wilson and R. B. Martin, *Arch. Biochem. Biophys., 142,* 445 (1971).

45. M. Greenberg and D. L. Rabenstein, unpublished results.

46. N. C. Li and R. A. Manning, *J. Amer. Chem. Soc.*, *77*, 5255 (1955).

47. R. B. Martin and J. T. Edsall, *J. Amer. Chem. Soc.*, *81*, 4044 (1959).

48. B. J. Fuhr and D. L. Rabenstein, *J. Amer. Chem. Soc.*, *95*, 6944 (1973).

49. G. K. R. Makar, M. L. D. Touche, and D. R. Williams, *J. Chem. Soc. Dalton,* 1016 (1976).

50. M. L. D. Touche and D. R. Williams, *J. Chem. Soc. Dalton,* 1355 (1976).

51. D. L. Rabenstein, *Can. J. Chem.*, *50*, 1036 (1972).

52. D. L. Rabenstein and S. Libich, *Inorg. Chem.*, *11*, 2960 (1972).

53. L.-O. Andersson, *J. Poly. Sci. A*, *10*, 1963 (1972).

54. G. D. Zegzhda, T. V. Zegzhda, and V. M. Shulman, *Russ. J. Inorg. Chem.*, *14*, 70 (1969).

55. W. Stricks and I. M. Kolthoff, *J. Amer. Chem. Soc.*, *75*, 5673 (1953).

56. R. C. Kapoor, G. Doughty, and G. Gorin, *Biochim. Biophys. Acta,* *100*, 376 (1965).

57. R. J. Kostelnik and A. A. Bothner-By, *J. Mag. Res.*, *14*, 141 (1974).

58. R. A. Haberkorn, L. Que, Jr., W. O. Gillum, R. H. Holm, C. S. Lin, and R. C. Lord, *Inorg. Chem.*, *15*, 2408 (1976).

59. B. Birgersson, R. E. Carter, and T. Drakenberg, *J. Mag. Res.*, *28*, 299 (1977).

60. J. Aaseth, *Acta Pharmacol. Toxicol.*, *39*, 289 (1976).

61. E. Ogawa, S. Suzuki and H. Tsuzuki, *Japan J. Pharmacol.*, *22*, 275 (1972).

62. A. M. Corrie and D. R. Williams, *J. Chem. Soc. Dalton,* 1068 (1976).

63. N. Taniguchi, T. Sato, T. Kondo, H. Tawachi, K. Saito, and E. Takakuwa, *Clinica Chimica Acta,* *59*, 29 (1975).

64. D. L. Rabenstein, S. Backs, and C. A. Evans, unpublished results.

65. T. L. Sayer, S. Backs, C. A. Evans, E. K. Millar, and D. L. Rabenstein, *Can. J. Chem.*, *55*, 3255 (1977).

66. G. A. Neville and T. Drakenberg, *Acta Chem. Scand. B,* *28*, 473 (1974).

67. D. L. Rabenstein and R. Saetre, *Anal. Chem.*, *49*, 1036 (1977).

68. D. L. Rabenstein and M. T. Fairhurst, *J. Amer. Chem. Soc., 97,* 2086 (1975).

69. D. L. Rabenstein and C. A. Evans, *Bioinorg. Chem., 8,* 107 (1978).

70. D. L. Rabenstein, *Acc. Chem. Res., 11,* 100 (1978).

71. R. B. Simpson, *J. Amer. Chem. Soc., 83,* 4711 (1961).

72. Y. Hojo, Y. Sugiura, and H. Tanaka, *J. Inorg. Nucl. Chem., 38,* 641 (1976).

73. R. Benesch and R. E. Benesch, *Arch. Biochem. Biophys., 38,* 425 (1952).

74. B. B. Hasinoff, N. B. Madsen, and O. Avramovic-Zikic, *Can. J. Biochem., 49,* 742 (1971).

75. B. J. White and R. G. Wolfe, *J. Chromatog., 7,* 516 (1962).

76. C.-C. Chin and J. C. Warren, *J. Biol. Chem., 243,* 5056 (1968).

77. J. Bloemmen and R. Lontie, *Biochim. Biophys. Acta, 236,* 487 (1971).

78. P. C. Leavis and S. S. Lehrer, *Biochemistry, 13,* 3042 (1974).

79. R. Cecil, *Biochem. J., 47,* 572 (1950).

80. R. E. Benesch and R. Benesch, *J. Amer. Chem. Soc., 75,* 4367 (1953).

81. R. E. Benesch and R. Benesch, *Arch. Biochem., 28,* 43 (1950).

82. J. H. Ladenson and W. C. Purdy, *Clin. Chem., 17,* 908 (1971).

83. J. W. Chang and R. B. Martin, *J. Phys. Chem., 73,* 4277 (1969).

84. B. Jezowska-Trzebiatowska, G. Formicka-Kozlowska, and H. Kozlowski, *Chem. Phys. Lett., 42,* 242 (1976).

85. J. E. Letter, Jr. and R. B. Jordan, *J. Amer. Chem. Soc., 97,* 2381 (1975).

86. R. B. Martin, M. Chamberlin, and J. T. Edsall, *J. Amer. Chem. Soc., 82,* 495 (1960).

87. S. T. Chow, C. A. McAuliffe, and B. J. Sayle, *J. Inorg. Nucl. Chem., 37,* 451 (1975).

88. H. Sakurai, A. Yokoyama, and H. Tanaka, *Chem. Pharm. Bull., 19,* 1416 (1971).

89. P. Hemmerich, in The Biochemistry of Copper (J. Peisach, P. Aisen, and W. E. Blumberg, eds.), Academic Press, New York, 1966, p. 15.

90. B. Jezowska-Trzebiatowska, G. Formicka-Kozlowska, and H. Kozlowski, *J. Inorg. Nucl. Chem., 39,* 1265 (1977).

91. W. Espersen and R. B. Martin, *J. Amer. Chem. Soc., 98,* 40 (1976).

92. J. Beattie, D. Fensom, and H. C. Freeman, *J. Amer. Chem. Soc.,* *98,* 500 (1976).

93. V. Vortisch, P. Kroneck, and P. Hemmerich, *J. Amer. Chem. Soc.,* *98,* 2821 (1976).

94. J. W. Dubnoff, *Biochem. Biophys. Res. Comm.,* *16,* 484 (1964).

95. F. Wagner and K. Bernhauer, *Ann. N.Y. Acad. Sci.,* *112,* 580 (1964).

96. D. Dolphin and A. J. Johnson, *J. Chem. Soc.,* 2174 (1965).

97. N. Adler, T. Medwick, and T. J. Poznanski, *J. Amer. Chem. Soc.,* *88,* 5018 (1966).

98. H. A. O. Hill, J. M. Pratt, R. G. Thorp, B. Ward, and R. J. P. Williams, *Biochem. J.,* *120,* 263 (1970).

99. G. Pellizer, G. R. Tauszik, and G. Costa, *J. Chem. Soc. Dalton,* 317 (1973).

100. P. Y. Law and J. M. Wood, *J. Amer. Chem. Soc.,* *95,* 914 (1973).

101. T. Frick, M. D. Francia, and J. M. Wood, *Biochim. Biophys. Acta,* *428,* 808 (1976).

102. S. Cockle, H. A. O. Hill, S. Ridsdale, and R. J. P. Williams, *J. Chem. Soc. Dalton,* 297 (1972).

103. T. J. Huang and G. P. Haight, Jr., *J. Chem. Soc., Chem. Comm.,* 985 (1969).

104. T. J. Huang and G. P. Haight, Jr., *J. Amer. Chem. Soc.,* *93,* 611 (1971).

105. S. G. Carr, P. D. W. Boyd, and T. D. Smith, *J. Chem. Soc. Dalton,* 907 (1972).

106. H. C. Freeman in Inorganic Biochemistry (G. L. Eichorn, ed.), Elsevier, New York, 1973, Vol. 1, p. 133.

107. D. Werner, S. A. Russell, and H. J. Evans, *Proc. Nat. Acad. Sci. U.S.,* *70,* 339 (1973).

108. R. Basosi, E. Tiezzi, and G. Valensin, *J. Phys. Chem.,* *79,* 1725 (1975).

109. T. R. Khan and C. H. Langford, *Can. J. Chem.,* *54,* 3192 (1976).

110. Y. Sugiura and H. Tanaka, *Biochem. Biophys. Res. Comm.,* *46,* 335 (1972).

111. N. S. Kosower and E. M. Kosower in Glutathione: Metabolism and Function (I. M. Arias and W. B. Jacoby, eds.), Raven Press, New York, 1976, p. 159.

112. P. J. Sadler, *Structure and Bonding,* *29,* 171 (1976).

113. A. A. Isab and P. J. Sadler, *J. Chem. Soc., Chem. Comm.,* 1051 (1976).

114. A. A. Isab and P. J. Sadler, manuscript in preparation.

115. A. McAuley and M. A. Olatunji, *Can. J. Chem., 55,* 335 (1977).

116. A. McAuley and M. A. Olatunji, *Can. J. Chem., 55,* 3328 (1977).

117. M. L. D. Touche and D. R. Williams, *J. Chem. Soc. Dalton,* 1355 (1976).

118. J. M. White, R. A. Manning, and N. C. Li, *J. Amer. Chem. Soc., 78,* 2367 (1956).

119. P. Kroneck, *J. Amer. Chem. Soc., 97,* 3839 (1975).

120. G. Oster and G. K. Oster, *Contraception, 10,* 273 (1974).

121. R. Cecil and J. R. McPhee, *Biochem. J., 66,* 538 (1957).

122. F. F. Brown, I. D. Campbell, P. W. Kuchel, and D. L. Rabenstein, *FEBS Lett., 82,* 12 (1977).

Chapter 5

COORDINATION CHEMISTRY OF L-DOPA
AND RELATED LIGANDS

A. Gergely and T. Kiss
Department of Inorganic and Analytical Chemistry
Kossuth Lajos University
Debrecen, Hungary

143

1. INTRODUCTION

Studies of the physiological and therapeutic effects of L-dopa and
related compounds (tyrosine, dopamine, noradrenaline, and adrenaline)
are at present in the foreground of biological, biochemical, and
pharmacological research. The question of the significance of the
roles of metal ions in the biological activities of these compounds
has recently arisen. Reviews have already appeared in this series,
for example, on the role of the formation of metal chelates in the
storage and transport of neurotransmitters (see Chap. 5, Vol. 6).
The biosyntheses of these compounds [1] are illustrated in Fig. 1.

Of these compounds, tyrosine is an amino acid; dopamine, noradrenaline, and adrenaline are catecholamines; and dopa, as a cate-
cholamino acid, displays similarities with both groups of compounds.

The catecholamines are hormones of the adrenal medulla, and
consequently exert very varied effects on the organism. They influence, among others, the blood circulation, the smooth muscles, and
the intermediate metabolism. Their effects on the organism and on
the individual organs coincide with the influence of a sympathetic
stimulus, from which it follows that these compounds are the chemical
substances of sympathetic nervous stimulus transmission. L-tyrosine
is a precursor of the catecholamines in the living organism. At the
same time, it is an important constituent of polypeptide metalloenzymes, in a considerable proportion of which the tyrosine residues
are the bonding sites of the metals [2].

The catecholamines find many different applications. For about
15 years L-dopa has been a very effective therapeutic in Parkinson's
disease [3], and in manganese poisoning, which is accompanied by
neurologically somewhat similar sequels [4].

Participation of the metalloenzymes in the metabolism of the
catecholamines is proved. For a better understanding of the physiological and biochemical reactions involved, therefore, it is necessary above all to know the details of the formation, composition,
stability conditions, and structure of the metal-catecholamine

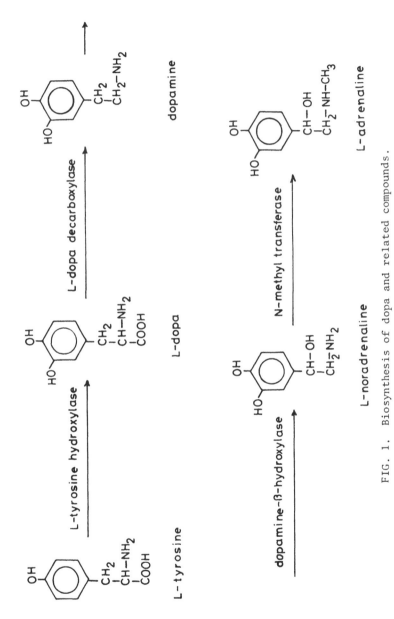

FIG. 1. Biosynthesis of dopa and related compounds.

complexes. In this chapter we deal first with the chemical properties of L-dopa and related compounds, then with their coordination
chemical behavior, and in part with the biological functions of the
metal complexes.

2. CHEMICAL CHARACTERISTICS OF L-DOPA AND RELATED LIGANDS

2.1. Acid-Base Characteristics

The dissociable protons of the ligands are to be found in the
following groups: in tyrosine in the carboxyl, the $-\overset{+}{N}H_3$, and
the phenolic hydroxy groups; in dopa in the carboxyl, the $-\overset{+}{N}H_3$,
and the two phenolic hydroxy groups; in dopamine in the $-\overset{+}{N}H_3$ and
the two phenolic hydroxy groups; in noradrenaline in the $-\overset{+}{N}H_3$,
the two phenolic hydroxy, and the one alcoholic hydroxy groups;
and in adrenaline in the $>\overset{+}{N}H_2$, the two phenolic hydroxy, and the
one alcoholic hydroxy groups (see Fig. 1). The deprotonation
macroconstants of these groups are listed in Table 1.

The acidities of the substituted ammonium and the first
phenolic hydroxy groups of these ligands are similar [9]. Hence,

TABLE 1

Macroconstants of Ligand Deprotonation

Ligand	pK_{COOH}	pK_1	pK_2	pK_{Ar-OH}	pK_{R-OH}	Ref.
tyr	2.20	9.03	10.08	-	-	5
dopa	2.22	8.80	9.83	13.4	-	6
dopam	-	8.89	10.41	13.0	-	7
nad	-	8.64	9.70	~13	~13	8
ad	-	8.66	9.95	~13	~13	8

Note: pK_1 + pK_2 together refer to the deprotonation of the $-\overset{+}{N}H_3$
and the first Ar-OH groups.

the pH metrically determined macroconstants arise as the superposition of the constants characteristic of the following microprocesses:

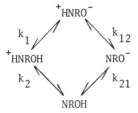

Edsall et al. [10] described a readily performable method for the complete elucidation of the deprotonation microprocesses of ligands of this type. Table 2 contains the microconstants obtained by this means, together with the values R = [$^+$HNRO$^-$]/[NROH] = k_1/k_2, the ratios of the concentrations of the zwitterionic and neutral forms of the ligands.

It is noteworthy that in the catecholamines (mainly in the case of adrenaline and noradrenaline) the acidity sequence of the -$\overset{+}{N}$H$_3$ and phenolic hydroxy groups is the reverse of that for tyrosine. The acidity of the first phenolic group in the case of the catecholamines is higher than that of tyrosine. This effect can be attributed in part to the hydrogen bonding occurring with the second phenolic hydroxy group [11]. The values of R = [$^+$HNRO$^-$]/[NROH] in Table 2

TABLE 2

Spectrophotometrically Determined Microconstants
of Ligand Deprotonation

Ligand	pk_1	pk_2	pk_{12}	pk_{21}	R = k_1/k_2	Ref.
tyr	9.63	9.28	9.69	10.04	0.44	10
dopa	8.93	9.19	9.63	9.43	1.8	12
dopam	8.87	9.95	10.36	9.39	12.0	7
nad	8.70	9.54	9.64	8.80	6.9	9
ad	8.72	9.57	9.89	9.04	7.1	9

also indicate that predominantly the zwitterion form is produced in the first dissociation process of the catecholamines.

The knowledge of the microconstants is important from both biological and coordination chemical aspects. If either the oxidation of the ligand (which is strongly pH-dependent) or the formation of metal complexes is the subject of study in kinetic investigations, incorrect conclusions may be reached if the acidities of the respective groups are not used. Although the use of the macroconstants is enough for a thermodynamic description of the formation of the metal complexes it is not sufficient for other conclusions to be drawn.

2.2. Oxidation Processes

In the presence of oxidizing agents in alkaline medium the catecholamines readily undergo oxidation. This process is appreciably accelerated by certain metal ions. The reactions can take place by several, still not completely clarified pathways. From a biological aspect the most important is the oxidation to aminochromes [13] and then to melanins [14]. The oxidation also proceeds at a lower pH (6 to 8) in the presence of metal ions, but the rate of the reaction and the end product depend on the buffer employed and on the metal ion catalyst itself. Strong catalytic effects are exerted by copper(II), manganese(III), vanadyl(II), and iron(III) ions whereas others, e.g., cobalt(II), nickel(II), were reported to inhibit oxidation [13].

The effects of complex-forming ligands on the metal ion-catalyzed oxidation of the catecholamines have not been clearly elucidated, but a few generalizations can be drawn. Compounds such as EDTA, CDTA, etc., which contain a large number of donor groups and which form very stable complexes with metal ions, are inhibitors. In contrast, those ligands which favor mixed ligand complex formation with the catecholamines in the presence of metal ions make the oxidation more effective than the metal ions themselves [15].

Pelizzetti et al. [16-20] have made a detailed study of the kinetics of the first step of the oxidation, the transformation to the o-quinone, in the presence of various metal ions. They assumed that the formation of outer sphere complexes between the metal ions and the phenolic hydroxy groups of the ligands plays a role in the mechanism of the reaction.

A wide-ranging study has been made of the oxidation of the catecholamines in the presence of a copper(II)-containing enzyme, tyrosinase [13]. In the living organism too this enzyme catalyzes the oxidation to the o-quinone, which is finally converted, through a series of reactions to the biologically important melanin. The mechanism of action of the enzyme has not been fully clarified. At any rate, it is certain that it involves the reversible redox transformation of the copper, and its direct interaction with the ortho phenolic hydroxy groups [21]. Yasunobu et al. [22] studied the oxidation of tyrosine-containing proteins and peptides in the presence of tyrosinase. They found that the N-terminal tyrosine peptides are oxidized in accordance with a "dopachrome" mechanism and the C-terminal tyrosine peptides in accordance with a "dopa-quinone" mechanism.

2.3. Ambivalent Properties

The ligands contain more than two donor groups, and thus in principle there are a number of possibilities for the formation of complexes of different bonding types. Although tyrosine contains three donor groups, for steric reasons only the amino acid side chain is suitable for chelate formation. Similarly, dopamine too contains only one, phenolic hydroxy chelate-forming donor group pair, and the participation of the side chain in the coordination is of subordinate importance. For tyrosine and dopamine, therefore, the ambivalent character is not manifested in practice.

Dopa, noradrenaline, and adrenaline each contain two separate chelate-forming groups within the molecule (see Fig. 1).
Via the orthophenolic hydroxy groups, all three ligands are capable of pyrocatechol-like O,O coordination, while via the side
chain dopa is able to coordinate in an amino acid-like N,O way,
and noradrenaline and adrenaline in an ethanolamine-like O,N way.
Depending on the metal ion, the pH, and the ligand, the formation
of mixed type complexes involving different donor groups is possible.
Because of the significant difference between the complex-forming
tendencies of the amino acids and ethanolamine, as well as its derivatives, it is to be expected that the ambivalent character of dopa will
be the most marked as a consequence of its alanine side chain. In the
cases of adrenaline and noradrenaline, however, it may be assumed that
the O,O bonding mode is favored, and that the participation of the
side chain in the complex formation is more subordinate. Since in
these three ligands the two chelate-forming donor groups are separated
within the molecule, there is the possibility that metal ions coordinate at both ends of the molecule, and thus also the possibility of
polynuclear polymeric complexes.

3. METAL COMPLEXES OF THE LIGANDS

In the case of tyrosine, extensive studies have been made of the
thermodynamic and kinetic conditions of complex formation with the
$3d^5$-$3d^{10}$ transition metal ions, and of the structural features of
the resulting complexes. Fairly abundant knowledge is available on
the complexes of the catecholamines with transition metal ions,
primarily copper(II), zinc(II), and nickel(II). Because of their
roles in the storage and transport of the neurotransmitters [23],
attention has also been focused on the study of the interaction
with calcium(II) and magnesium(II) ions.

3.1. Nickel, Copper, and Zinc Complexes

3.1.1. *Complexes with Tyrosine*

The bidentate nature of tyrosine has been proved in many ways [24-27]. Gergely and Kiss [25] carried out pH metric and calorimetric studies in the pH range 3 to 11. It was found that in the case of nickel(II) complexes with compositions of 1:3 are formed. The thermodynamic data on the complexes were calculated by considering the monoprotonated form of tyrosine as the chelating agent (Table 3).

The equilibrium data on the complexes MHA^+, $M(HA)_2$, and $M(HA)_3^-$ display agreement with the corresponding constants of other bidentate amino acids. The first deprotonation constants of the complexes approximately coincide with the corresponding microconstant (pk_{21}) of the ligand (see Table 2), whereas the subsequent constants are larger in accordance with statistical considerations. Similarly, the deprotonation heats too exhibit an approximate agreement with the deprotonation micro enthalpy of the ligand ($\Delta H_1 = 23.1$ kJ mol^{-1}). The data for the zinc(II) complexes, however, indicate that the deprotonation of the phenolic hydroxy group and the hydrolysis of the complexes occur together.

3.1.2. *Complexes with Dopa*

Dopa (see Fig. 1) is potentially a tetradentate ligand, and thus the possibility exists for the formation of N,O complexes with the alanine side chain, and O,O complexes with the ortho phenolic hydroxy groups. Investigations with metal-alanine-pyrocatechol systems containing the potential donor groups of dopa on separate ligands [6, 12] have provided information on the properties of nickel(II), copper(II), and zinc(II) complexes of dopa. It was established that the hardest of these ions, zinc(II), favors O,O coordination, the least hard, nickel(II), favors N,O coordination, while, depending on the pH, copper(II) tends to form either type of bonding. However, the formation of complexes containing both N,O and O,O bonds is favored for all three metal ions.

TABLE 3

Thermodynamic Data of Nickel(II)-Tyrosine, Copper(II)-Tyrosine, and Zinc(II)-Tyrosine Complexes*

Process	Nickel(II)			Copper(II)			Zinc(II)		
	log K	ΔH [kJ M^{-1}]	ΔS [J deg^{-1}]	log K	ΔH [kJ M^{-1}]	ΔS [J deg^{-1}]	log K	ΔH [kJ M^{-1}]	ΔS [J deg^{-1}]
$M^{2+} + HA^- \rightleftharpoons MHA^+$	4.90	-12.8	49	7.69	-22.0	72	4.12	-9.2	47
$MHA^+ + HA^- \rightleftharpoons M(HA)_2$	4.28	-18.1	20	6.75	-30.7	24	3.83	-10.8	36
$M(HA)_2 + HA^- \rightleftharpoons M(HA)_3^-$	2.87	-18.8	-9	--	--	--	--	--	--
$MHA^+ \rightleftharpoons MA + H^+$	--	--	--	-9.32	23.3	-98	-8.14	37.5	-28
$M(HA)_2 \rightleftharpoons MHA_2^- + H^+$	-9.32	24.0	-95	-10.06	22.6	-115	-8.89	29.7	-68
$MHA_2^- \rightleftharpoons MA_2^{2-} + H^+$	-9.92	25.9	-101	--	--	--	-9.29	40.0	-41
$M(HA)_3^- \rightleftharpoons MH_2A_3^{2-} + H^+$	-9.04	26.1	-83	--	--	--	--	--	--
$MH_2A_3^{2-} \rightleftharpoons MHA_3^{3-} + H^+$	-10.41	25.9	-110	--	--	--	--	--	--
$MHA_3^{3-} \rightleftharpoons MA_3^{4-} + H^+$	-10.06	25.5	-105	--	--	--	--	--	--

*I = 0.2 M KCl; t = 25° [25].

Equilibrium Constants of Nickel(II)-Dopa, Copper(II)-Dopa and Zinc(II)-Dopa Complexes

Process	Ni(II) Ref. 12[a]	Ni(II) Ref. 30[b]	Cu(II) Ref. 6[c]	Cu(II) Ref. 28[d]	Zn(II) Ref. 12[a]	Zn(II) Ref. 30[b]
$\log K_1$ $M^{2+} + H_2A^- \rightleftharpoons MH_2A^+$	4.90	4.96	7.52	7.12	3.77	4.1[e]
$\log K_2$ $MH_2A^+ + H_2A^- \rightleftharpoons M(H_2A)_2$	3.70	4.20	6.63	6.29	--	--
$\log K_1^*$ $M^{2+} + HA^{2-} \rightleftharpoons MHA$	10.3	--	--	12.99	10.6	9.94
$\log K_2^*$ $MHA + HA^{2-} \rightleftharpoons M(HA)_2^{2-}$	--	--	--	11.95	--	8.12
$-\log K^H$ $MH_2A^+ \rightleftharpoons MHA + H^+$	8.23	--	--	--	6.77	--
$-\log K_1^H$ $M(H_2A)_2 \rightleftharpoons MH_3A_2^- + H^+$	7.87	--	6.80	--	--	--
$-\log K_2^H$ $MH_3A_2^- \rightleftharpoons MH_2A_2^{2-} + H^+$	9.00	--	8.47	--	8.59	--
$-\log K_3^H$ $MH_2A_2^{2-} \rightleftharpoons MHA_2^{3-} + H^+$	9.62	--	9.51	--	9.67	--
$-\log K_4^H$ $MHA_2^{3-} \rightleftharpoons MA_2^{4-} + H^+$	11.47	--	10.35	--	10.42	--
$\log K_p$ $MH_2A^+ + MA^- \rightleftharpoons M_2A_2H_2$	--	--	15.08	14.61	--	--
$\log K_c$ $M_2H_2A_2 + MA^- \rightleftharpoons M_2A_2^{2-} + MH_2A^+$	--	--	11.77	12.43[f]	--	--

[a],[c] $I = 0.2$ M KCl; 25°.

[b],[d] $I = 1.0$ M KNO$_3$; 25°.

[e] Ref. 31; $I = 0.5$ M NaNO$_3$; 20°.

[f] Ref. 29; $I = 1.0$ M KNO$_3$; 25°.

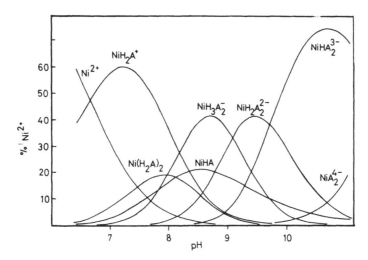

FIG. 2. Concentration distribution of complexes formed in the nickel(II)-dopa system as functions of pH at a metal ion-to-ligand ratio of 1:2, $C_M = 2 \times 10^{-3}$ M, 0.2 M KCl, 25°.

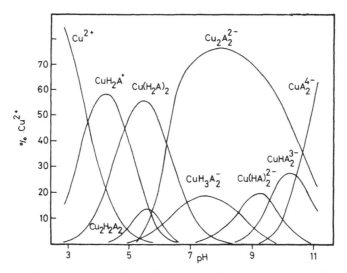

FIG. 3. Concentration distributions of complexes formed in the copper(II)-dopa system as functions of pH at a metal ion-to-ligand ratio of 1:2, $C_M = 2 \times 10^{-3}$ M, 0.2 M KCl, 25°.

The equilibrium and in part structural studies by Gorton and Jameson [28, 29] on the copper(II)-dopa system supported the formation of 1:1 and 1:2 amino acid-like and pyrocatechol-like complexes when there is a ligand excess. At a metal ion-to-ligand ratio of 1:1, they concluded that polymers are formed. It was assumed that they are built up of species with a 1:1 composition, containing N,O bonds. In addition to these, the formation of a cyclic dimeric complex (i.e., containing two units) was also described. Regarding the nickel(II)-dopa system they concluded [30] that only amino acid-like 1:1 and 1:2 complexes are formed. In the case of the zinc(II)-dopa system, the formation of O,O-bonded species is considered more favored.

Equilibrium studies of the nickel(II)-dopa, copper(II)-dopa, and zinc(II)-dopa systems in the pH interval 3 to 11 have also been made by Gergely et al. [6, 12]. It was concluded that the complexes with the following stoichiometric compositions are formed: NiH_2A^+, $Ni(H_2A)_2$, $NiHA$, $NiH_3A_2^-$, $Ni(HA)_2^{2-}$, $NiHA_2^{3-}$, NiA_2^{4-}; CuH_2A^+, $Cu(H_2A)_2$, $CuH_3A_2^-$, $Cu(HA)_2^{2-}$, $CuHA_2^{3-}$, CuA_2^{4-}, $Cu_2H_2A_2$, $Cu_2A_2^{2-}$; ZnH_2A^+, $ZnHA$, $ZnH_3A_2^-$, $Zn(HA)_2^{2-}$, $ZnHA_2^{3-}$, ZnA_2^{4-}.

The equilibrium constants for the respective complex formation are given in Table 4. The concentration distributions of the species formed in these systems at a metal ion-to-dopa ratio of 1:2 are illustrated in Figs. 2-4.

The existence of 1:3 complexes was not demonstrated in the nickel(II)-dopa system [12, 30]. This was partially explained by the steric hindrance caused by the intramolecular hydrogen bonding arising between the noncoordinated phenolic hydroxy groups and the donor groups of the side chain [12].

In the nickel(II)-dopa and zinc(II)-dopa systems, there is overlapping of the processes of formation of metal complexes containing N,O- and O,O-type bonding and the dissociation processes of the donor groups not bound to the metal ion. With regard to the copper(II) complexes the formation of the complexes containing the different donor groups takes place in seaparted intervals. The

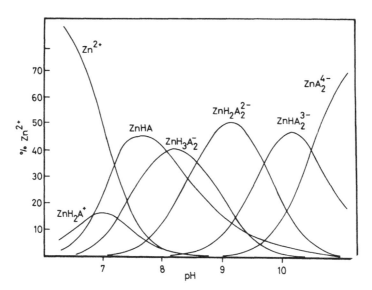

FIG. 4. Concentration distributions of complexes formed in the zinc(II)-dopa system as functions of pH at a metal ion-to-ligand ratio of 1:2, $C_M = 2 \times 10^{-3}$ M, 0.2 M KCl, 25°.

processes of complex formation and of deprotonation of the donor groups are similarly better differentiated from one another [6].

In efforts to clarify the bonding modes in the complexes formed, visible [6, 11, 28, 32], UV [12], CD [11], and ESR (electron spin resonance) [6, 33, 34] spectral procedures have been employed. Supported by the results of these experiments too, we shall now discuss in detail the considerations relating to the bonding types of the complexes formed:

(1) With nickel(II) or copper(II) as central metal ion, the amino acid-type complexes of compositions MH_2A^+ and $M(H_2A)_2$ are formed in significant concentrations (see Figs. 2 and 3), whereas in the case of zinc(II) only the complex ZnH_2A^+ is formed (see Fig. 4).

The CD, visible spectral, and cyclic voltametric studies [11, 32], proved that at a metal ion-to-dopa ratio of 1:2, up to pH ~ 5.4, only N,O-bonded complexes are formed in the copper(II)-dopa system. Kustin et al. [35, 36] assumed that in the case of amino acid-type dopa complexes of copper(II) and nickel(II) one of the phenolic

hydroxy groups is linked to the metal ion via a hydrogen bond with
a water molecule. In connection with these conclusions, Boggess and
Martin [11] drew attention to the error caused by the use of macro-
protonation constants not characteristic of the individual donor
groups.

(2) On the basis of the magnitude of the deprotonation constants $\log K^H$ the bonding mode of NiHA and ZnHA is probably pyrocatechol-like [12, 34]. In the copper(II)-dopa system, when ESR spectral
studies were carried out the presence of CuHA could not be confirmed
[6].

(3) A mixed complex-like species containing both N,O and
O,O bonding (structure I) is strongly suggested for the complex

(I)

The complex $MH_3A_2^-$ of dopa.

$MH_3A_2^-$ [12]. This bonding mode appears to be confirmed by visible
spectral studies in the nickel(II)-dopa system, and by ESR and
visible spectral studies in the copper(II)-dopa system. Boggess
and Martin [11] observed the appearance of a charge-transfer band
at 436 nm and a d-d transition at 635 nm at copper(II)-dopa-OH⁻
ratios of 1:2:3 (pH ~ 7.6) and 1:1:3 (pH ~ 9) in the copper(II)-
dopa system. On this basis, they likewise concluded that the
coordination involved three O and one N donor atoms.

(4) Three structures can be ascribed to the species $MH_2A_2^{2-}$.
Since there is not a significant difference in the corresponding
microconstants (pk_{12} and pk_{21}) of the ligand, in the case of a
mixed complex bonding mode, two different structures can be assumed
(structures II and III). In addition, deprotonation of the species

(II)   (III)

(IV)

Possible structures of the complex $MH_2A_2^{2-}$ of dopa.

$MH_3A_2^-$ may also be accompanied by rearrangement to the O,O bonding mode (structure IV). This is also supported by the $-\log K_2^H$ values which cannot be ascribed unambiguously to the dissociation of the hydroxy or the $\overset{+}{N}H_3$ groups. In the case of copper(II), the ESR spectral data [6] indicate that the structural rearrangement proceeds almost completely. This assumption is strengthened by Kwik et al. [32] who isolated the ammonium salt of this complex. In the case of zinc(II) [12], a UV spectral examination provided evidence that the structural rearrangement is not complete. Consequently, it may be assumed that the species corresponding to structures II, III, and IV are all present in equilibrium with one another. The visible and near-infrared spectra of the nickel(II)-dopa system, from pH ~ 9.5 undergo practically no change and agree well with the spectra of the N,O and O,O bound mixed complex Ni(alaninate)-(tiron)$^{3-}$ [12]. Accordingly, rearrangement to the pyrocatechol-like structure does not take place to a significant extent even at higher pH. In agreement with these, the $-\log K_2^H$ values vary in the sequence copper(II) < zinc(II) < nickel(II), in a manner analogous to the extent of the rearrangement to the pyrocatechol-like structure.

(5) The $-\log K_3^H$ value is approximately the same for all three metal ions and corresponds to the microconstant (pk_{12} in Table 2) for the dissociation of the $-\overset{+}{N}H_3$ group. Spectral studies on the complex $NiHA_2^{3-}$ strongly indicate the mixed bonding mode (structure V). In the case of zinc(II), however, the UV spectral data [12] confirm that rearrangement to the O,O bonding mode is complete, and thus the structure VI can be ascribed to the complex $ZnHA_2^{3-}$. This same structure also applies to the species $CuHA_2^{3-}$.

(V)   (VI)

Possible structures of the complex MHA_2^{3-} of dopa.

(6) The values of the constants $-\log K_4^H$ are almost the same in the cases of the copper(II) and zinc(II) complexes, and can be ascribed to the dissociation of the $-\overset{+}{N}H_3$ group of the structure VI species. The nickel(II) complex, on the other hand, remains of mixed bonding type, and thus the further proton-releasing process can be interpreted by the formation of a hydroxo complex [12].

(7) Polymeric species too exist in appreciable concentration in the copper(II)-dopa system. Formation of the cyclic dimeric complex of composition $Cu_2A_2^{2-}$ (structure VII) assumed by Gorton and Jameson [29] was confirmed, even in the case of ligand excess, by the visible spectral studies by Boggess and Martin [11]. Evidence was also provided that in this complex the ligands are linked to the metal ion by mixed-like bonding. As a result of ESR spectral

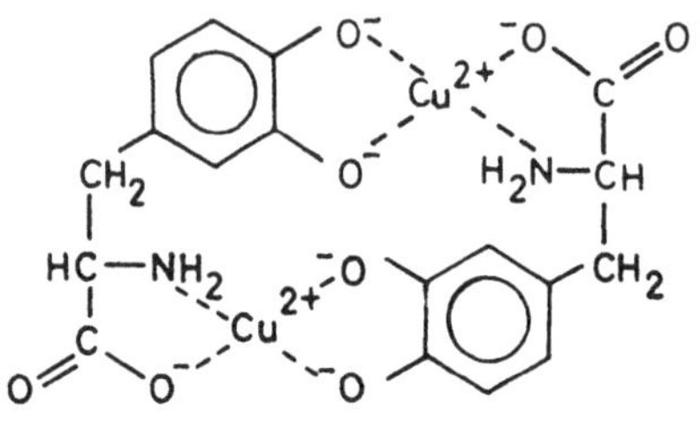

(VII)

Cyclic binuclear complex $Cu_2A_2^{2-}$ of dopa.

measurements at room temperature, Pilbrow et al. [33] exclude the
formation of the dimeric species. At the same time, in frozen solutions at 77 K, at copper(II)-dopa ratios of both 1:1 and 1:2 the
ESR spectrum is indicative of the presence of a dimeric species
[33, 34]. On the basis of results recorded in frozen solutions,
and the unusually shaped, broadened, almost symmetrical signal
obtained at room temperature, Gergely and Kiss [12] were of the
opinion that a dimeric species is formed at room temperature too.
The lack of change in the intensity of the signal was interpreted
in terms of the triplet state of the copper(II).

Gorton and Jameson [29] and Gergely and Kiss [6] further
assumed the formation of a similarly mixed bound chain-like polymeric complex (structure VIII). This complex is presumably formed

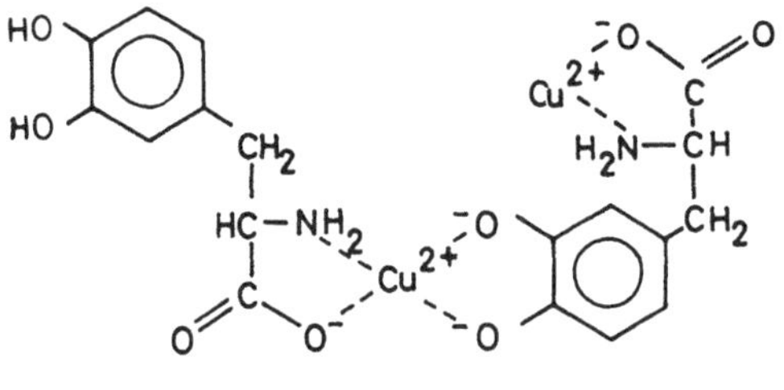

(VIII)

Chain-like polynuclear Cu^{2+} complex of dopa.

by the coupling of two MH_2A^+ monomers containing N,O bonds and is
accompanied by the formation of an O,O bond. Therefore, this formation process is written in such a way in Table 4 (log K_p) that the
deprotonated form of MH_2A^+, MA^-, appears in the equation.

3.1.3. Complexes with Dopamine

Dopamine contains only one chelate-forming donor group pair. Weber et al. [31] carried out pH metric investigations on the copper(II)-dopamine and nickel(II)-dopamine systems. From their experiments, they concluded that the complexes $CuHA^+$, CuA, $Cu(HA)_2$, CuA_2^{2-}, and $NiHA^+$, $Ni(HA)_2$ and NiA_2^{2-} are formed. Weber et al. [31] studied the zinc(II)-dopamine system only up to pH 7.6. On the basis of the equilibrium examinations, and taking into consideration the previous [31] findings too, Kiss and Gergely [7] assumed the formation of the species MHA, MA, $M(HA)_2$, MHA_2^-, and MA_2^{2-} for all three metal ions. (In the case of zinc(II), the complex ZnA^{2-} is not formed.) The stability constants and the deprotonation constants of the complexes are listed in Table 5.

The constants $\log K_1^*$ and $\log K_2^*$ approximately agree with the stability data for the complexes of pyrocatechol. The deprotonation constants for the complex $M(HA)_2$ correspond to the microconstant of the ligand (see Table 2). This suggests that dopamine is coordinated to these metals only via the phenolic hydroxy groups. The binding

TABLE 5

Equilibrium Constants of Nickel(II)-Dopamine,
Copper(II)-Dopamine and Zinc(II)-Dopamine Complexes[*]

Process		Ni(II)	Cu(II)	Zn(II)
$M^{2+} + HA^- \rightleftharpoons MHA^+$	$\log K_1^*$	9.42	14.27	10.26
$MHA^+ + HA^- \rightleftharpoons M(HA)_2$	$\log K_2^*$	6.34	11.66	8.77
$MHA^+ \rightleftharpoons MA + H^+$	$-\log K^H$	9.96	7.6	--
$M(HA)_2 \rightleftharpoons MHA_2^- + H^+$	$-\log K_1^H$	10.05	10.17	10.26
$MHA_2^- \rightleftharpoons MA_2^{2-} + H^+$	$-\log K_2^H$	10.81	10.62	10.88
$\log K_1^*/K_2^*$		3.08	2.61	1.49

[*]I = 0.2 M, KCl; t = 25° [7].

of the second ligand is sterically hindered in the case of nickel(II)
(log K_1^*/K_2^* = 3.08), and therefore deprotonation of the species NiHA$^+$
is favored. In the case of the copper(II) complex, on the other
hand, the possibility for deprotonation is given by the fact that
the electronic transition metal ion ← O results in an electronic
shift extending to the side chain too, and hence the acidity of the
chain terminal -$\overset{+}{N}H_3$ group increases considerably [7].

3.1.4. Complexes with Adrenaline and Noradrenaline

Similarly to dopa, adrenaline and noradrenaline also contain two
separate chelate-forming groups (see Fig. 1). The stabilities of
the O,O-bonded complexes are substantially higher than those of
the ethanolamine-like O,N complexes, and thus the complex-forming
properties of the side chain may be manifested only slightly.
Jameson and Neillie have made a study of the complexes of these
ligands with copper(II) [37] and, among others, nickel(II) and
zinc(II) [38]. In every case they assumed formation mostly of the
O,O-bonded complexes.

       With copper(II) or nickel(II) as the central ion, at low con-
centration the existence of complexes of mixed bonding type (structure
IX) and of complexes containing only O,N bonding was not excluded.

(IX)

Mixed bonding-type complexes of
adrenaline and noradrenaline.

In addition to this, as to the nickel(II) complexes a surprising
difference was observed in the complex-forming properties of
adrenaline and noradrenaline. This was interpreted in that, with
the formation of the 1:1 complex, the secondary amino group of the

adrenaline molecule becomes much more basic than the primary amino group of the noradrenaline. By this means the complex-forming role of the side chain increases, and thus the possibility arises for the formation of the mixed bonding complex (structure IX).

The occurrence of polymeric species was concluded at a metal ion-to-ligand ratio of 1:1 in both the copper(II)-adrenaline and copper(II)-noradrenaline systems. The metal ion forms O,N bonds with the side chains of the ligands, and a tetranuclear cyclic polymeric species is produced. In this way, the resulting tetramer consists of units containing only O,O bonding or only O,N bonding. On the other hand, the CD and visible spectral studies by Boggess and Martin [11] indicate a mixed complex-like bonding mode, which can be assumed on analogy with the dimeric complex of dopa.

3.2. Complexes of Other Metals

The number of publications relating to other metal complexes of the ligands is comparatively low. Even then, some of these papers yield only approximate or qualitative findings. Therefore, we can rely in some cases only on the partial similarities with the metal ions discussed in the preceding section.

Results relating to the manganese(II), iron(II), and cobalt(II) complexes of tyrosine [39, 40] similarly support that tyrosine behaves as a bidentate ligand. It is generally valid that the stability constants of the processes of $M^{2+} + A^- \rightleftharpoons MA^+$ and $MA^+ + A^- \rightleftharpoons MA_2$ are almost the same. They are relatively unstable and readily undergo hydrolysis. When the results obtained for the complexes of cadmium(II) and mercury(II) with tyrosine are compared with those for the zinc(II) complexes, it may be stated that with these three metal ions both 1:1 and 1:2 chelates are formed, but owing to the different symmetry of the mercury(II) complexes binding of the second ligand is less favored.

From a comparison of the experimental data reported for the iron(II) [41], iron(III) [42], cobalt(II) [38], manganese(II) [38],

and chromium(III) [43] complexes of dopa and the catecholamines, it
may be stated that in the event of a ligand excess only the phenolic
hydroxy groups take part in the complex formation. The ambivalent
nature of the ligands is displayed only at a metal ion-to-ligand
ratio of 1:1, or possibly in the case of a metal ion excess. Under
such conditions mononuclear and polymeric species of mixed bonding
type may be formed.

With regard to the complexes of dopamine, adrenaline, and
noradrenaline with cadmium(II) and lead(II) ions, it is known merely
that 0,0-bonded complexes are formed, whereas also amino acid-like,
N,0-bonded complexes are produced between these metal ions and dopa
[32].

The investigations by Gorton and Jameson [30] into the biologically important calcium(II) and magnesium(II) complexes led to
quantitative findings. They concluded that in the magnesium(II)-
dopa system a pyrocatechol-like 1:1 complex is formed. Rajan et al.
[44] also support 0,0 coordination in the calcium(II) and mag-
nesium(II) complexes of noradrenaline and dopamine, but at a higher
ligand excess they assume the binding of a second and a third ligand
molecule too.

3.3. Mixed Ligand Complexes

From a complex chemical aspect the systems containing a catecholamine
and a metal ion are in themselves very complicated, and this is probably one of the reasons why comparatively few results have been published on the ternary systems.

In connection with the mixed ligand complexes of tyrosine with
other amino acids and peptides, we only refer here to two reviews
which have already appeared in the present series [45, 46]. Since
tyrosyl residues act as metal-binding sites in a number of metalloenzymes, the in vitro study of the mixed complexes of tyrosine may
serve as a model for the investigation of the metalloenzyme-substrate
interaction.

Kiss and Gergely [7] made a study of the mixed complexes of nickel(II), copper(II), or zinc(II) ions with dopamine (which forms only pyrocatechol-like complexes) and alanine or pyrocatechol. It was found that the metal ion-dopamine-pyrocatechol mixed complexes, containing only 0,0 bonding, are formed to an extent corresponding to the statistical case. At the same time, an increased stability was observed in the case of the mixed complexes formed with the N,O-bonded alanine. This was interpreted via the different stability relations of the parent complexes with dopamine and alanine, and via the role of the charge neutralization. It is noteworthy that the mixed complex is the predominant species in the physiological pH interval, which again suggests that formation of the mixed complex may be of importance in the biochemical reactions of the catechol-amines.

The combined presence of significant quantities of iron(II), zinc(II), copper(II), magnesium(II), and calcium(II) ions, adenosine triphosphate (ATP), and biogenic amines (among them dopamine, nora-drenaline, and adrenaline) in vesicles of the adrenal medulla and of the terminal varicosities of sympathetic nerves permits the conclu-sion that the amines may form complexes with the metal ions and with ATP in the process of either their storage or their biochemical, physiological, or pharmacological action [1, 23]. Rajan et al. [47, 48] investigated systems containing the above-mentioned metal ions, ATP, and catecholamines. They concluded that the metal ion-ATP com-plexes are formed at lower pH (pH 3-6.5), and 1:1:1 metal ion-ATP-amine mixed complexes above pH 6.5. Further, the formation of water-soluble polymeric complexes was also strongly suggested. The inter-actions of the magnesium(II), cobalt(II), and nickel(II) ions with adrenaline and ATP have been studied by other authors with an nmr method too [49], and it was concluded that the following interactions exist in the mixed ligand complexes of cobalt(II) and of magnesium(II):

$$\text{adrenaline} \underline{\hspace{1cm}} M^{2+} \underline{\hspace{1cm}} \text{ATP}$$

where full lines denote strong bonds and dashed lines weak
interactions. It was concluded that the adrenaline participates in 0,0 coordination, and that the phosphate groups of ATP
are bound to the metal ion. The weak bonding between the ligands
was interpreted as an interaction between the adenine ring of ATP
and the catechol ring of adrenaline. Seifter et al. [50] even
succeeded in isolating a magnesium(II)-ATP-adrenaline complex, and
assumptions were made as to its mixed-complex structure. Nuclear
magnetic resonance examinations by Granot and Fiat [51] led to a
partially different result in connection with the interactions
between the metal ions, ATP, and catecholamines. In their view,
two types of metal ion-nucleotide-catecholamine complexes may be
formed from adrenaline and ATP, or from dopamine and ATP, in the
presence of magnesium(II) or cobalt(II). In one of these complexes
the fully protonated form of the catecholamine is linked to the
nucleotide without any direct bonding to the metal ion. In the
other the dissociated catecholamine molecule is bound to the metal
ion directly. In both species the nucleotide is strongly bonded to
the metal ion. In their view the concentration of the latter type
complex is negligible in the physiological pH range. The linkage of
the catecholamine molecules to ATP in the mixed complex does not
differ substantially from the metal ion-free catecholamine-ATP interaction.

4. IMPLICATION OF METAL IONS AND THEIR COMPLEXES IN BIOLOGICAL FUNCTIONS OF CATECHOLAMINES

The storage and transport of catecholamines has been dealt with in
detail by Rajan et al. in the present series (Vol. 6, Chap. 5) [23].
It may be expected that a quantitative study of the parent and
mixed ligand complexes discussed in Sec. 3, a wider ranging proof
of the structural assumptions, and an examination of the cell components and the role of the storage sites, will make further contributions to the better understanding of this subject.

The rate-determining step in the metabolism of the catecholamines is the conversion of tyrosine to dopa on the action of the iron-containing tyrosine hydroxylase. In the partially still hypothetical mechanism of the hydroxylation reaction it is assumed that an enzyme-Fe(III)-O_2^- complex is formed, which in effect acts as hydroxylating agent [21]. The exact mechanism of how the activity of this enzyme increases so dramatically from one instant to the next on the action of a stimulus has by no means been clarified. It is a noteworthy finding, however, that certain divalent metal ions, e.g., calcium(II), activate tyrosine hydroxylase also directly [52]. In the knowledge of the fact that the catecholamines inhibit the functioning of the enzyme, and that calcium(II) forms stable complexes with both the substrate and the inhibitor catecholamines, the question arises of the importance of the role of complex formation in the activation of the enzyme.

Another metalloenzyme-catalyzed process in the metabolism of the catecholamines is the hydroxylation of dopamine to noradrenaline on the action of the copper-containing dopamine-β-hydroxylase. In the mechanism of the hydroxylation, which is accompanied by the reversible redox transformation copper(II) ⇌ copper(I), the participation of an as yet unidentified complex, enzyme-$[Cu(I)]_n$-O, is strongly suspected [21]. At the same time, Goldstein [53] came to the conclusion from epr and nmr investigations that the dopamine neither changes the valency state of the copper, nor enters into a direct interaction with it.

According to the partially still hypothetical mechanism of the oxidation of the catecholamines as catalyzed by the copper-containing enzyme tyrosinase, there is a direct linkage between the copper(II) and the phenolic hydroxy groups of the substrate [21]. Harrison et al. [54] reached the opinion that there is an interaction between the copper(II) and the side chain of the substrate molecule too. These authors observed stereospecificity in the tyrosinase-catalyzed oxidation of the optical isomers of dopa, adrenaline, and noradrenaline. The rate of the initial step of the oxidation varied in the

sequence L-dopa > D-dopa > L-ad > D-ad > L-nad > D-nad. On the basis
of the stereospecificity and the dependence on the nature of the side
chain, it was concluded that tyrosinase contains a stereospecific
reaction rate control site. They assumed that the side chain of the
catecholamines is linked to the control site by forming a chelate,
via the copper, and this acts as a regulator of the reaction rate.

It appears that the role of the side chain is more pronounced
in the metal ion-catalyzed oxidation of the catecholamines to aminochrome. Gorton and Jameson [37] made a study of the copper(II) complexes of adrenaline and noradrenaline. They observed that in the
case of a ligand excess when there is a possibility for the formation
of nonautooxidizable complexes containing only 0,0 bonding there is
no aminochrome formation, whereas at a metal ion-to-ligand ratio of
1:1 when the complex-forming property of the side chain too is manifested the oxidation proceeds even when air is carefully excluded.
On this basis, it was assumed that the tetrameric complex formed at
a metal ion-to-ligand ratio of 1:1 is a transitional product in the
copper(II)-catalyzed formation of the aminochromes.

5. CONCLUSIONS

The complex formation properties of dopa and related compounds,
primarily with metal ions of biological importance, have been reviewed. It is a general finding, that tyrosine is capable only of
N,0 bonding and dopamine only of 0,0 bonding. Dopa, noradrenaline,
and adrenaline, which contain two separate chelate-forming groups,
can coordinate to a metal ion both via their ortho phenolic hydroxy
groups and via the donor groups of the side chain. Which donor
group pair takes part predominantly in the formation of a coordinate
bond is determined by the strengths of the bonds between the metal
ion and the donor group pairs (as the primary factor), but also by
the pH and the metal ion-ligand ratio.

The roles of the metal ions and of the metal ion-catecholamine
interactions in certain biological, physiological, and biochemical

processes of the ligands have been touched on. One possibility for answers to be obtained to the many as yet unsolved questions is the further in vitro study of the interaction with metal ions of the catecholamines and the other potential ligands participating in their reactions in the living organism. This might contribute to a fuller understanding of both their reactions in biological systems and their widespread therapeutic application and effect.

ABBREVIATIONS

ad	adrenaline
Ar-OH	phenolic hydroxy group
CDTA	1,2-cyclohexanediaminetetraacetate
dopa	3,4-dihydroxyphenylalanine
dopam	dopamine
EDTA	ethylenediaminetetraacetate
K_c	formation constant of the dimeric complex (see structure VII)
K_p	formation constant of the polymer (see structure VIII)
nad	noradrenaline
R-OH	alcoholic hydroxy group
tyr	tyrosine
pK_{Ar-OH}	deprotonation macroconstant of the second phenolic hydroxy group

REFERENCES

1. J. M. Musacchio, in Handbook of Psychopharmacology, Vol. 3, Biochemistry of Biogenic Amines (L. L. Iversen, S. D. Iversen, and S. H. Snyder, eds.), Plenum Press, New York and London, 1975, p. 1.

2. B. L. Vallee and W. E. C. Wacker, in The Proteins, Vol. 5, Metalloproteins (H. Neurath, ed.), Academic Press, New York and London, 1970, p. 54.

3. O. Hornykiewicz, *Life Sci.*, *15*, 1249 (1974).

4. J. Mena, J. Court, S. Fuenzalida, P. S. Papavasilion, and G. C. Cotzias, *New Eng. J. Med.*, *282*, 5 (1970).

5. A. Gergely, I. Nagypál, T. Kiss, and R. Király, *Acta Chim. Acad. Sci. Hung.*, *82*, 257 (1974).

6. A. Gergely and T. Kiss, *Inorg. Chim. Acta*, *16*, 51 (1976).

7. T. Kiss and A. Gergely, *Inorg. Chim. Acta*, to be published.

8. R. F. Jameson and W. F. S. Neillie, *J. Chem. Soc.*, 2391 (1965).

9. R. B. Martin, *J. Phys. Chem.*, *75*, 2657 (1971).

10. J. T. Edsall, R. B. Martin, and B. R. Hollingworth, *Proc. Nat. Acad. Sci. U.S.*, *44*, 505 (1958).

11. R. K. Boggess and R. B. Martin, *J. Amer. Chem. Soc.*, *97*, 3076 (1975).

12. A. Gergely, T. Kiss, and Gy. Deák, *Inorg. Chim. Acta*, to be published.

13. R. A. Heacock, *Chem. Rev.*, *59*, 181 (1959); and *Adv. Heterocyclic Chem.*, *5*, 205 (1965).

14. G. A. Swan, *Ann. N.Y. Acad. Sci.*, *100*, 1005 (1963).

15. G. Losse, A. Barth, and W. Langenbeck, *Chem. Ber.*, *94*, 2271 (1961).

16. E. Pelizzetti, E. Mentasti, and G. Girandi, *Inorg. Chim. Acta*, *15*, L1 (1975).

17. E. Pelizzetti, E. Mentasti, and E. Pramanro, *J. Chem. Soc. Dalton*, 23 (1976).

18. E. Mentasti, E. Pelizzetti, and C. Baiocchi, *J. Inorg. Nucl. Chem.*, *38*, 2017 (1976).

19. E. Pelizzetti, E. Mentasti, E. Pramanro, and M. E. Carlotti, *Gazz. Chim. Ital.*, *105*, 307 (1975).

20. E. Pelizzetti, E. Mentasti, and G. Saini, *Gazz. Chim. Ital.*, *106*, 605 (1976).

21. El-Ichiro Ochiayi, *J. Inorg. Nucl. Chem.*, *37*, 1503 (1975).

22. K. T. Yasunobu, E. W. Peterson, and H. S. Mason, *J. Biol. Chem.*, *234*, 3291 (1959).

23. K. S. Rajan, R. W. Colburn, and J. M. Davis, in Metal Ions in Biological Systems, Vol. 6 (H. Sigel, ed.), Marcel Dekker, New York and Basel, 1976, Chap. 5.

24. J. E. Letter and J. E. Bauman, *J. Amer. Chem. Soc.*, *92*, 443 (1970).

25. A. Gergely and T. Kiss, unpublished results.

26. O. A. Weber, *J. Inorg. Nucl. Chem.*, *36*, 1341 (1974).

27. M. L. Barr, E. Baumgartner, and K. Kustin, *J. Coord. Chem.*, *2*, 263 (1973).

28. J. E. Gorton and R. F. Jameson, *J. Chem. Soc.*, *A*, 2616 (1968).

29. J. E. Gorton and R. F. Jameson, *J. Chem. Soc. Dalton*, 304 (1972).

30. J. E. Gorton and R. F. Jameson, *J. Chem. Soc. Dalton*, 310 (1972).

31. B. Grgas-Kužnar, Vl. Simeon, and O. A. Weber, *J. Inorg. Nucl. Chem.*, *36*, 2151 (1974).

32. W.-L. Kwik, E. Purdy, and E. J. Stiefel, *J. Amer. Chem. Soc.*, *96*, 1638 (1974).

33. J. R. Pilbrow, S. G. Carr, and T. D. Smith, *J. Chem. Soc.*, *A*, 723 (1970).

34. S. G. Carr, T. D. Smith, and J. R. Pilbrow, *J. Chem. Soc.*, *A*, 2569 (1971).

35. R. L. Karpel, K. Kustin, A. Kowalak, and R. F. Pasternack, *J. Amer. Chem. Soc.*, *93*, 1085 (1971).

36. M. L. Barr, K. Kustin, and Sung-Tsuen Lin, *Inorg. Chem.*, *12*, 1486 (1973).

37. R. F. Jameson and W. F. S. Neillie, *J. Inorg. Nucl. Chem.*, *27*, 2623 (1965).

38. R. F. Jameson and W. F. S. Neillie, *J. Inorg. Nucl. Chem.*, *28*, 2667 (1966).

39. A. Ya. Suchev and P. K. Mihaly, *Biokhimiya*, *27*, 25 (1962).

40. A. Gergely, I. Nagypál, and B. Király, *Acta Chim. Acad. Sci. Hung.*, *68*, 285 (1971).

41. G. Litwack, *Life Sci.*, 509 (1962).

42. R. Abu-Eittah, Z. Mobarak, and S. El-Lathy, *J. Prakt. Chem.*, *316*, 235 (1974).

43. Yu. P. Davidov, N. I. Voronik, and A. V. Skriptsoba, *Radio-khimiya*, 597 (1975).

44. K. S. Rajan, J. M. Davis, and R. W. Colburn, *J. Neurochem.*, *18*, 345 (1971).

45. R. P. Martin, M. M. Petit-Ramel, and J. P. Scharff, in Metal Ions in Biological Systems, Vol. 2 (H. Sigel, ed.), Marcel Dekker, New York, 1973, Chap. 1.

46. D. D. Perrin and R. P. Agarwal, in Metal Ions in Biological Systems, Vol. 2 (H. Sigel, ed.), Marcel Dekker, New York, 1973, Chap. 4.

47. K. S. Rajan, J. M. Davis, R. W. Colburn, and F. H. Jarke, *J. Neurochem.*, *19*, 1099 (1972).

48. K. S. Rajan and J. M. Davis, *J. Inorg. Nucl. Chem.*, *38,* 897 (1976).

49. I. Muro, I. Morishima, and T. Yonezawa, *Chem.-Biol. Interactions,* *3,* 213 (1971).

50. J. Seifter, E. Seifter, and G. Guideri, *Amer. J. Med. Sci,* *263,* 261 (1972).

51. J. Granot and D. Fiat, *J. Amer. Chem. Soc.,* *99,* 4963 (1977).

52. V. H. Morgenroth, M. C. Boadle-Biber, and R. H. Roth, *Mol. Pharmacol.,* *11,* 427 (1975).

53. M. Goldstein, in The Biochemistry of Copper (J. Peisach, P. Aisen, and W. E. Blumberg, eds), Academic Press, New York and London, 1966, p. 443.

54. W. H. Harrison, W. W. Whisler, and S. Ko, *J. Biol. Chem.,* *242,* 1660 (1967).

Chapter 6

STEREOSELECTIVITY IN THE METAL COMPLEXES
OF AMINO ACIDS AND DIPEPTIDES

Leslie D. Pettit and Robert J. W. Hefford
Department of Inorganic and Structural Chemistry
The University
Leeds, England

173

1. INTRODUCTION

Isomerism in the formation of metal complexes has been a subject of
study since the days of Werner, and the realization that ligand molecules could adopt various conformations increased the range of potential isomers considerably. When the ligand contains one or more
asymmetric centers the number of possible isomers increases yet
again, and provides a field of research in itself. This field has
been reviewed by Bernauer [1], who covered much of the literature
until about 1973.

Many biological systems that involve metal ions are remarkably
selective in the chirality of the ligand molecules with which they
interact; many are even stereospecific. However, understanding of
this stereospecificity is disappointingly limited. The only reliable
results so far have come from model compounds and the simpler molecules, such as amino acids and dipeptides, that are present in biological systems. Using such small molecules it has been shown
quantitatively that ligands with differing chiralities do, in fact,
behave differently. However, the differences are generally smaller
than those found in the biological field.

This review covers some recent results in the field of thermodynamic stereoselectivity in the complexes of amino acids and dipeptides. The emphasis will be on α-amino acids important in the biological field and on dipeptides prepared from these basic units.
Some synthetic-substituted amino acids will also be included but
polycarboxylic acids and similar ligands will not be considered.

1.1. Definitions

If one takes the formation of a bis-amino acid complex of a metal
ion M with an optically active amino acid HL as an illustrative
example, three possible species can result, $[M(+L)_2]$, $[M(-L)_2]$,
and $[M(+L)(-L)]$, in addition to any geometrical or conformational
isomers. The first two complexes will have identical stabilities

while the last complex, containing ligands of opposite chiralities, is diastereoisomeric (a meso complex) and may well have a different stability.

If it is assumed that there is no intrinsic stabilization or destabilization of the meso species as a result of different atomic interactions, the equilibrium mixture of M with racemic amino acids would be expected to contain the bis complexes in the statistically expected ratios of

$$[M(+L)_2] : [M(-L)_2] : [M(+L)(-L)] = 1:1:2$$

As a result the equilibrium constant K for the ligand exchange reaction:

$$\tfrac{1}{2}[M(+L)_2] + \tfrac{1}{2}[M(-L)_2] \rightleftharpoons [M(+L)(-L)] \qquad (1)$$

has the value K = 2, and

$$2\beta_{M(+L)_2} = 2\beta_{M(-L)_2} = \beta_{M(+L)(-L)}$$

that is

$$\log \beta_{M(+L)(-L)} = \log \beta_{M(+L)_2} + 0.30 \qquad (2)$$

Thermodynamic stereoselectivity, expressed as $\Delta \log \beta$, may be detected quantitatively by deviations from this relationship, that is

$$\Delta \log \beta = \log \beta_{M(+L)_2} - \log \beta_{M(+L)(-L)} + 0.30 \qquad (3)$$

Stereoselectivity is therefore positive when the optically active bis complex is favored. It should be noted that, as a result of relationship (3), stereoselectivity can still be positive even though the species $[M(+L)_2]$ is apparently less stable than the species $[M(+L)(-L)]$.

The macroconstants calculated from normal potentiometric titrations cannot distinguish between these various chiral bis complexes, just as they cannot distinguish between cis and trans isomers, but it can be shown [2] that:

$$\beta_2(\text{macro}) = 0.5\,\beta_{M(+L)_2} + 0.25\,\beta_{M(+L)(-L)}$$

As a result, in the absence of stereoselectivity the titration curve
of a metal ion in the presence of a racemic ligand would not be expected to differ from that of the optically pure ligand. It follows
that, in the extreme case of complete stereospecificity (i.e.,
$\beta_{M(+L)(-L)} = 0$), values for log β_2 (macro) calculated from the
racemic mixture will be 0.30 log units lower than from the optically
pure ligand. With dipeptides comparison is generally made between
complexes of the optically active dipeptides (e.g., L-Val-L-Val) and
the meso isomer (L-Val-D-Val). The most important complexes are
always the mono species and the optically active and meso dipeptide
can be obtained pure. Hence there are no complications from statistical factors.

The measurement of formation constants can, in general, only
be applied reliably to the formation of labile complexes. It is
quite likely that kinetic stereoselectivity in the formation of
inert complexes may result in one particular optical isomer being
formed in preference to another as a result of differing activation
energies. In some cases (e.g., with Co(III)) almost complete
stereoselectivity can result [3]. However, in biological systems
labile complexes generally predominate with the result that thermodynamic stereoselectivity is the more important aspect.

1.2. Origins of Stereoselectivity

Thermodynamic stereoselectivity can result from differences in
either entropy or enthalpy changes or both. It is possible to
account for statistical factors which will influence the entropy
changes [2] but other factors may also contribute to give anomalous
values for ΔS. These include:

1. Changes in the coordination number that depend on the
   chirality of the ligand involved
2. Changes in the complex-solvent interface (often described
   as hydrophobic interactions) and other factors that will
   influence the entropy of solvation

3. Changes in vibrational entropy of the complexes as a
   result of conformational differences

Some of these factors will be discussed in more detail later.

Stereoselectivity in the enthalpy changes accompanying complex
formation would be expected to be more significant in many cases.
Chelate rings are often puckered and, when puckered, substituent
groups on atoms in the ring can adopt either equatorial or axial
conformations, with the more bulky substituents preferring equatorial
positions. The diamine chelate ring has been most extensively
studied in this connection. In general amino acid chelate rings are
less puckered, hence such steric interactions would not be expected
to be so important. However many amino acids contain side chains
which can either interact with the metal ion or can disturb the
complex-solvent interface. If it is assumed that the amino acid
bonds glycine-like in the first instance, substituent groups can
influence further coordination in a number of interrelated ways,
such as:

(1) If one assumes planer bis coordination as an intermediate
step in complex formation, cis and trans isomers are possible. If
both amino acids are of the same chirality both side chains will be
on the same side of the complex in the trans isomer, and on opposite
sides in the cis complex. In the mesocomplex the reverse will be
true (see Fig. 1). If the side chain can coordinate with the metal
ion along the z axis then, in the optically active bis complex, only
one side chain can interact with the metal in the trans isomer while
both can coordinate in the cis isomer. Hence stereochemical control

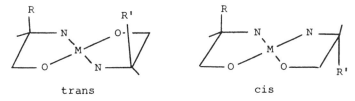

trans   cis

FIG. 1

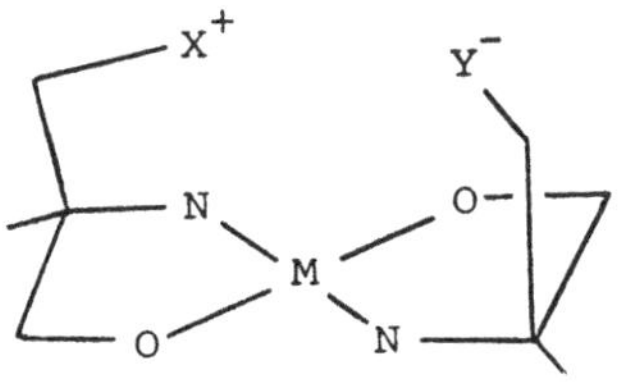

FIG. 2

of the geometrical isomers formed is to be expected. Alternatively, if there is a significant difference in stability between the cis and trans isomers a significant difference is to be expected in the stabilities of the meso and optically active complexes $[M(+L)(-L)]$ and $[M(+L)_2]$. In addition, the formation of a tris complex will be discouraged when both ligands in the bis complexes are sterically able to behave as tridentate ligands. Tris complex formation will be encouraged when steric constraints discourage tridentate coordination.

(2) If the side chains in ternary bis complexes can interact with one another then, again, the cis-trans equilibrium will be disturbed. For example, if the side chains are of opposite charge, electrostatic interaction (often hydrogen bonding) will favor the isomer in which both substituent groups are on the same side of the complex as shown in Fig. 2.

(3) The side chains may not necessarily interact to form conventional bonds to one another or to the metal ion for stereoselectivity to result. It is found that hydrophobic side chains tend to prefer to be concentrated on the same side of the complex although no classical bonding can be envisaged. Such hydrophobic interaction will result in changes in solvation and will have a large entropy contribution but some enthalpy component will also be present. In some complexes substituent groups on the side chains tend to "stack" one above the other. This will clearly be influenced by the stereochemistry of the component ligands.

(4) A type of nonclassical interaction takes place between some transition metal ions and aromatic ring systems in the amino

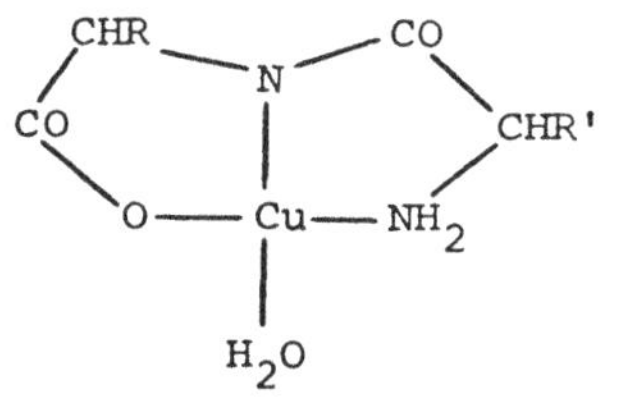

FIG. 3

acid side chain (e.g., phenylalanine or tryptophan). This interaction is supported by thermodynamic measurements [4], and X-ray
diffraction studies of crystals [5, 6]. When present such metal-
aromatic interactions will give the ligand a tridentate character
so that stereoselectivity can follow as described in (1).

With dipeptide complexes of Cu(II) the major species are those
in which the dipeptide behaves as a tridentate ligand, bonding
through the amine nitrogen, the amide nitrogen, and a carboxyl
oxygen (the [M(L-1H)] or 11-1 species). The ligand is therefore
held in a planar conformation as in Fig. 3 and the substituent
groups will orientate themselves either both on the same side of
the complex (the optically active isomer) or on opposite sides (the
meso isomer). In practice it is found that substituent groups have
far more importance in complexes of dipeptides than in those of
amino acids.

1.3. Model Compounds

Amino acids and dipeptides are themselves small molecules of biological importance and so can act as models for more complicated
systems. However, the diamine chelate ring is considerably more
puckered than that of amino acids [7] and hence can give information
on the influence of axial and equatorial substituents that cannot
be detected in amino acid complexes. When inert complexes are
formed, the conformation of a substituent group as small as the
methyl group can so influence the course of the reaction that

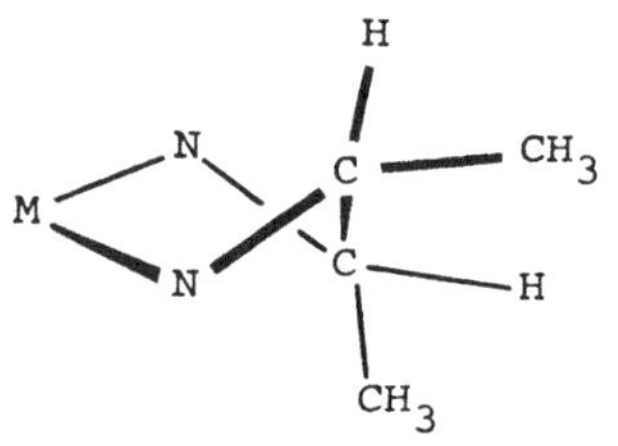

FIG. 4

virtually complete stereoselectivity results [3]. With labile complexes, differences in stability between the various conformations
can be calculated [7, 8] but, in practice, they have little influence
on the thermodynamics of complex formation when the substituent groups
are small [9]. When the diamine contains two asymmetric centers
(e.g., 2,3-diaminobutane, bn) both optically active and meso isomers
exist. It is impossible for both substituent groups to occupy equatorial positions in the chelate ring formed by the meso ligand when
the ring is in the preferred puckered conformation (see Fig. 4). As
a result both mono and bis complexes of meso-bn are significantly
less stable than those of (±)bn, largely as a result of less favorable
enthalpy changes [10]. With potentially octahedral metal ions such
as Ni(II) or Zn(II) coordination differences also contribute. With
1,3 diamines more conformation isomers are possible and stereoselectivity has been detected by contact shift measurements [11].

Diamines contain two nitrogen donor atoms. Tartaric acid has
only oxygen donor atoms. Tartaric acid exists as both optically
active and meso isomers and stereoselectivity in the formation of
tartrate complexes has been known for a considerable time [12]. The
situation is further complicated by polynuclear complex formation,
particularly with Cu^{2+} and VO^{2+}, but the stereoselectivity found in
both acid dissociation constants and in binuclear complex formation
with VO^{2+} has recently been explained as resulting from conformational differences [13].

1.4. Detection of Stereoselectivity

Techniques for detecting stereoselectivity have been reviewed previously in this series [14]. Spectroscopic methods have been widely
quoted but there are always problems in the interpretation of results
which, in general, can only lead to qualitative conclusions. Potentiometric titrations, supported where possible by calorimetric studies,
provide the more reliable quantitative results. If titrations of a
labile metal ion in the presence of an optically pure amino acid are
compared with those using a racemic mixture of the ligand, super-
imposable curves signify the absence of thermodynamic stereoselectivity [2]. Total stereospecificity in favor of the optically active
bis complex will cause the calculated values for $\log \beta_2$ (macro) to
differ by 0.30 log units since optically active bis complexes can
still be formed from the racemic ligand. This will mean that pH
differences will always be less than 0.3 pH units. Hence the precision of the extent of stereoselectivity, even under ideal circumstances, cannot be expected to be better than ±5% and will often be
significantly worse.

With dipeptides, comparison is generally between formation
constants calculated from separate titrations of the two isomerically
pure ligands (e.g., L-Val-L-Val and L-Val-D-Val) as mentioned in
Sec. 1.1. The measurement of stereoselectivity between these diastereoisomers will not suffer from the above limitation and will
therefore be generally more precise.

The use of ligand exchange chromatography has enabled stereoselectivity to be detected in systems where, in some cases, the
extent is too small to be found potentiometrically [15-17]. However,
it is difficult to understand the results quantitatively. The use
of ligand exchange chromatography to resolve optical isomers of
ligands is an important field for research and has many potential
applications, some of which will be discussed later. The chirality

of an amino acid coordinated to an inert metal ion may have a pro-
found effect on the course of a subsequent coordination reaction
[18, 19]. Exchange chromatography may then be used to separate the
isomers formed during the reactions [18].

2. AMINO ACID COMPLEXES

α-Amino acids of general formula $^+NH_3 \cdot CHR \cdot CO_2^-$ (HL) may conveniently
be classified according to the nature of the side chain substituent
group R. For the purpose of this review the following rather arbi-
trary divisions will be made:

1. Essentially noncoordinating side chains. These may be
   either aliphatic or aromatic.
2. Side chains containing potentially coordinating "hard"
   donor centers. These may have only very limited donor
   capabilities (e.g., serine) or may be effective donors
   (e.g., glutamic acid).
3. Side chains containing "soft" donor centers. These may
   be sulfur atoms (e.g., penicillamine) or aromatic nitrogen
   donors (e.g., histidine).

2.1. Noncoordinating Side Chains

If one ignores the effects of hydrophobic interactions and potential
metal-aromatic ring interactions, amino acids in this group all bond
glycine-like. The only effects the substituents can have will be
through their inductive effects (not stereoselective) or through
steric interactions (possibly stereoselective). As a result small
substituent groups would not be expected to show significant stereoselectivity, and the influence would be expected to be greater in
peptide complexes than in amino acid complexes because, in the
former, the ligand is held more rigidly.

Both potentiometric and spectrophotometric techniques have
been used in the study of binary complexes of simple amino acids and
the majority of the investigations have failed to detect stereoselectivity. Wellman et al. concluded that little interaction exists
between the two amino acid ligands in the bis complex [20]. Gillard
et al. failed to find stereoselectivity in copper(II) complexes of
alanine, valine, and proline [21], and the work was extended to complexes of phenylalanine, tyrosine, and tryptophan by Simeon and
Weber, with similar results [22]. On the other hand, electronic
absorption spectra of $[Cu(ValO)_2]$ and $[Cu(ProO)_2]$ have been interpreted as indicating some stereoselectivity [23], although this is
contrary to the findings of most other investigations. More recent
results have all generally agreed with the above conclusions.

It is interesting to note that, although ligands such as
phenylalanine and tryptophan contain aromatic substituent groups
that appear to interact with metal ions, they do not promote significant stereoselectivity in simple binary complexes. However,
there is evidence that in some ternary complexes these substituent
groups can promote stereoselectivity [24]. These will be discussed
in Sec. 2.3.

Although stereoselectivity appears to be absent, or very small,
in naturally occurring bidentate amino acids, large substituent
groups such as N-benzyl make it significant, particularly in ternary
complexes. Using N-benzyl-L-proline (L-BzPro), Davankov and Mitchell
showed that $[Cu(L-BzProO)(D-ProO)]$ is more stable than $[Cu(L-BzProO)$
$(L-ProO)]$, with a similar result being probable with valine [23].
However, more recently Davankov et al. studied the complexes of
Cu(II) with N-benzyl-D/L-valine and found that the chiral complexes,
$[Cu(L-BzValO)_2]$ and $[Cu(D-BzValO)_2]$, were more stable than meso
$[Cu(L-BzValO)(D-BzValO)]$. They attributed this to a destabilization
of the meso complex as a result of interligand interactions [25].
They also investigated the effect of changing the solvent on the
stereoselectivity in complexes of Cu(II) with some N-substituted

amino acids using spectrometric and polarimetric methods [26]. They
found that changing the solvent resulted not only in a change in the
magnitude of the stereoselectivity but could even cause a change in
its sign.

Snyder and Angelici studied a wide range of N-carboxymethyl
and N-benzyl-N-carboxymethyl amino acids [27], extending the work
of Leach and Angelici [28]. They found that, in general, the ternary
copper complex containing ligands of the same optical hand was more
stable than its meso isomer. However, there were some exceptions,
particularly N-carboxy-methyl-L-aspartic acid and -L-glutamic acid,
which tended to prefer a D-amino acid as the ternary ligand. Both
aspartic and glutamic acids are tridentate amino acids and are
discussed in Sec. 2.2.

Stereoselectivity has also been found in the bis complexes of
N-(o-hydroxybenzylalanine) with Co(II) and Ni(II), whereas it is
absent with Cu(II) and Zn(II) [29]. With both Co(II) and Ni(II)
the stereoselectivity is in favor of the optically active bis complexes, stabilization being about 3 kJ mol^{-1}. It was shown that
this preference could be the result of space crowding, which is
minimized in one particular isomer of the optically active bis
complex.

When the labile complex [Cu(N-carboxymethyl-L-ValO)]$^{+}$ is
bonded chemically to a styrene-divinylbenzene copolymer, the resulting exchange resin resolves partially valine, leucine, proline, and
alanine [16]. It was found that the degree of resolution increased
as the bulkiness of the side chains increased, and it appears that
the polystyrene resin matrix does more than simply bind the amino
acid derivative; it also enhances the stereoselectivity of the
complex [27]. This is in agreement with the general conclusion that
N substitution, particularly with aromatice groups, enhances stereoselectivity.

2.2. Side Chains Containing Hard Donor Centers

Amino acids falling within this group range from those with only weakly coordinating groups, such as serine (R = CH_2OH), threonine [R = $CH(CH_3)OH$], and tyrosine (R = $CH_2 \cdot C_6H_4OH$), to those with reactive coordination sites, such as glutamic acid [R = $(CH_2)_2CO_2H$] and lysine [R = $(CH_2)_4NH_2$].

There is some doubt as to whether the hydroxy groups of ligands such as serine have significant donor properties prior to ionization of the proton. With amino acids this does not take place until a high pH (>10) has been reached, even in the presence of metal ions, and hence can be disregarded in the context of this review. In dipeptides, it has been found that ionization of tyrosyl phenolic protons can lead to some interesting dimers with Cu(II). Serine and threonine are both biologically important amino acids that interact strongly with both Ni(II) and Cu(II). For example, the ternary complex [Cu(L-HisO)(L-ThrO)] accounts for a significant proportion of the complexed labile Cu(II) in human serum and has therefore been studied extensively. In spite of this the extent of coordination through the hydroxyl oxygen is uncertain. In the solid state the hydroxyl group does not coordinate [30]. In solutions, however, the ternary complex is significantly more stable than expected from the stabilities of the parent complexes [31]. Stereoselectivity in formation of the ternary complex was found to be insignificant [32].

Although stereoselectivity appears to be absent in the formation constants of [Cu(SerO)$_2$] and [Cu(ThrO)$_2$], a detailed calorimetric study has detected a small but significant difference in the enthalpies of formation of [Cu(L-SerO)$_2$] and [Cu(L-SerO)(D-SerO)], the optically active isomer having a bond energy greater by 1.7 ± 0.2 kJ mol^{-1} [33]. This stereoselectivity was opposed by a negative stereoselectivity in the entropy change with the result that the difference in formation constants was not significant. Stereoselectivity was explained by outer sphere coordination between Cu(II)

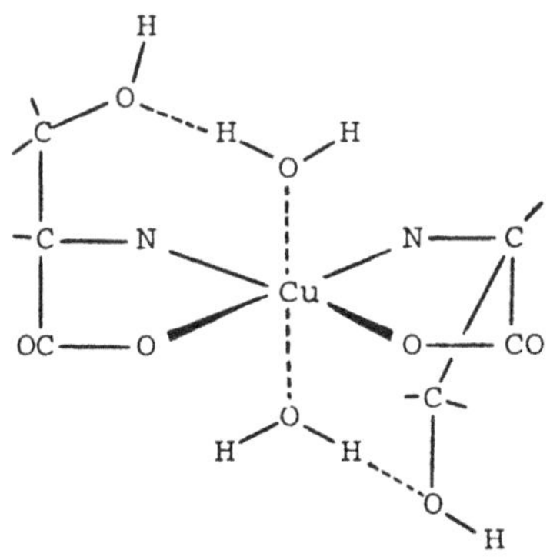

FIG. 5

and the hydroxyl oxygens, as shown in Fig. 5. To adopt this structure the serine molecules would have to coordinate cis with the
serinate ion in a gauche conformation as shown in Fig. 5. Such a
conformation is found in crystals of $[Cu(SerO)_2]$ [34].

Measurement of formation constants has also failed to detect
stereoselectivity in the formation of parent binary complexes of
amino acids with reactive side chains containing hard donor centers.
Ritsma et al. studied complexes of asparagine (R = CH_2CONH_2), glutamine [R = $(CH_2)_2CONH_2$], glutamic acid [R = $(CH_2)_2CO_2H$], and aspartic
acid (R = CH_2CO_2H) with a number of transition metal ions and were
unable to detect stereoselectivity [35]. A similar absence was found
with 2,3-diaminopropionic acid [36], and this general trend has been
confirmed in most later studies.

When present in ternary complexes, however, ligands in this
class can promote stereoselectivity, particularly when N-substituted
[27]. For example, when coordinated to inert metal ions, the chirality of aspartic acid [37] and asparagine [18] will dictate the course
of reactions with additional ligands such as the oxalate ion or
diaminoethane. The mixed ligand Co(III) complexes of L-aspartic-N-
monoacetic acid with various simple glycine-like amino acids have
been shown to form the cis-N and trans-N isomers stereospecifically
[19].

Some applications of stereoselectivity will be described later. One of these applications involves the optical resolution of DL-aspartic and DL-glutamic acids via the formation of ternary complexes with Cu(II) and L-arginine, [R = $(CH_2)_3NH \cdot C(NH)NH_2$], L-lysine and L-ornithine [R = $(CH_2)_3NH_2$] [38]. This stereoselectivity was attributed to ligand-ligand interactions in the ternary complexes, resulting in the preferential formation of the meso complex.

2.3. Side Chains Containing Soft Donor Centers

2.3.1. Complexes of Ligands Containing Sulfur

"Soft" is a rather subjective word when used to describe donor properties. In the context of this review it will be used to describe the heavy donor atom, sulfur, in amino acids such as methionine [R = $(CH_2)_2SCH_3$] and penicillamine [R = $C(CH_3)_2SH$], and pyridine-like nitrogen donor atoms as in histidine (Fig. 7) and its analogs.

Thioether sulfur atoms are found in methionine, S-methylcysteine (R = CH_2SCH_3), and lanthionine ($S[CH_2CH(NH_2)CO_2H]_2$). Methionine and S-methylcysteine have been studied a number of times but there are no certain conclusions concerning the presence or absence of sulfur bonding to Cu(II) or Ni(II). Early solution studies were interpreted as showing only glycine-like bonding [39] and this was supported by a solid state crystal structure determination [40]. However, other evidence is less specific and proton magnetic resonance (PMR) spectra suggest that, whereas metal-S bonding is apparently absent with methionine, it is present with S-methylcysteine [41]. Potentiometric studies with Cu(II) failed to detect any stereoselectivity, as would be expected with glycine-like coordination. However, a small effect was found with Ni(II), the meso-bis complex being marginally more stable than the optically active bis complex [Ni(L-MetO)$_2$] [42]. A calorimetric study of the Ni(II)-methionine system suggested that this stereoselectivity was a result of enthalpy rather than entropy differences [42]. It was concluded that tridentate coordination was probably present. Hence the bonding was more than simply "glycine-like."

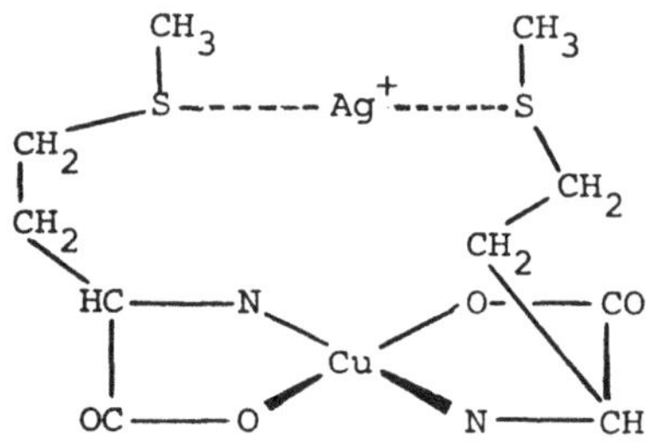

FIG. 6

An interesting example of stereoselectivity in complexes of methionine with copper is found in the binuclear complex $[CuAg(MetO)_2]^+$ shown in Fig. 6 [43]. The silver ion can be bonded between the two sulfur atoms to form a discrete molecule (rather than a polymer) if the original Cu(II) bis complex contains two L-methionate ions bonded trans, or one L- and one D-bonded cis. The binuclear complex $[CuAg(L-MetO)_2]^+$ was found to be the more stable, suggesting coordination of methionate around the Cu(II) to be trans [44]. This is the geometry found in crystalline $[Cu(MetO)_2]$ [40].

Lanthionine $(S[CH_2CH(NH_2)CO_2H]_2)$ forms complexes with Cu(II) in which there is good evidence for Cu-S bonding [45]. Because the amino acid contains two asymmetric centers, stereoselectivity between complexes of the meso and optically active forms is to be anticipated. In fact, the meso isomer was found to be the more stable [45].

Cysteine $(R = CH_2SH)$ and penicillamine $[R = C(CH_3)_2SH]$ both contain thiol sulfur atoms. As a result they are readily oxidized and so more difficult to study. However, the Ni(II) and Zn(II) complexes of D/L-penicillamine (Pen) and D/L-cysteine have been studied by Ritsma and Jellinek [46]. No stereoselectivity was observed for the cysteine complexes and that between Zn(II) and penicillamine was on the edge of significance in favor of the meso complex. Significant stereoselectivity was found in the Ni(II)-penicillamine complexes, the meso complex [Ni(L-penO)(D-penO)] being more stable, with a Gibbs free energy difference of 2.3 ± 0.2 kJ mol^{-1}.

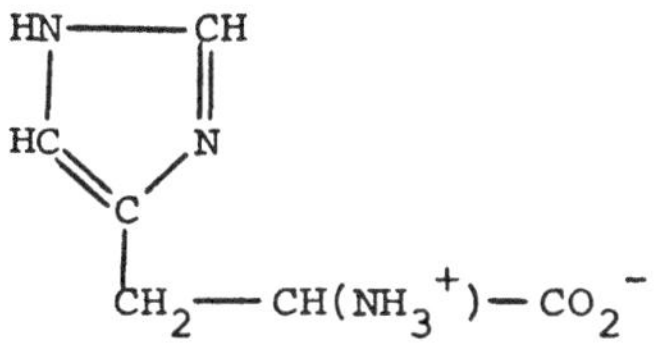

FIG. 7

2.3.2. *Complexes of Histidine*

Stereoselectivity is most significant in complexes of histidine
(Fig. 7) and its substituted analogs. As a result these species
have been studied in more detail than any others.

The first evidence for the stereoselectivity of histidine came
from an nmr study of Co(II)-histidine complexes [47]. It was found
that the meso complex [Co(D-HisO)(L-HisO)] was more stable than the
chiral species, with a value of K = 5.8 for the reaction:

$$0.5[Co(L-HisO)_2] + 0.5[Co(D-HisO)_2] \rightleftharpoons [Co(L-HisO)(D-HisO)]$$

(4)

The statistically expected value would be 2. Hence a "true" stabilization of about 1.3 kJ mol^{-1} was found. This system, together with
the Ni(II)-histidine system, was studied potentiometrically by Ritsma
et al. [48]. They confirmed the earlier observation finding comparable values of 1.3 kJ mol^{-1} for the Co(II)-histidine system and
2.1 kJ mol^{-1} for the Ni(II)-histidine system. Simultaneously, Morris
and Martin studied the complexes of histidine with Co(II), Ni(II),
Zn(II), Cd(II), and Cu(II) [36]. They confirmed the above results
for Co(II) and Ni(II), found comparable stereoselectivity with Zn(II)
[ΔG (corrected) for equilibrium (4) = 1.3 kJ mol^{-1}], but were unable
to detect significant stereoselectivity with Cd(II) or Cu(II).

A combined potentiometric and calorimetric study of the bis-
histidinate complexes of Ni(II), Zn(II), Cu(II) [49], and Co(II) [50]
has permitted a better understanding of the origins of thermodynamic
stereoselectivity. The results, corrected for statistical factors

TABLE 1

Thermodynamic Values for Reaction (4), Corrected
for the Statistical Factor of RT ln 2, at 25°
and I = 0.10 M (kJ mol^{-1})

	ΔG-RT ln 2	ΔH	TΔS-RT ln 2
Co(II)	-1.70	-1.39	+0.3 (±0.2)
Ni(II)	-2.73	-2.55	+0.2
Cu(II)	0	+2.05	+2.0
Zn(II)	-0.68	-2.48	-1.8

to make them comparable to those quoted above, are shown in Table 1.
These results show that the stereoselectivity found with Co(II) and
Ni(II) is almost entirely due to different enthalpy changes, the
corrected difference in entropy changes being insignificant. With
Cu(II) the apparent absence of stereoselectivity is found to result
from a coincidental cancelling of enthalpy and entropy differences,
the optically active bis complex, $[Cu(L-HisO)_2]$, being preferred
from the point of view of bond energies. Hence the stereoselectivity
in ΔH for Cu(II) is of an opposite sign to that found with the other
transition metal ions. The absence of an entropy contribution to
the stereoselectivity of complexes of Co(II) and Ni(II) is reasonable
because coordination of both water and the histidinate ion would be
octahedral, and stereoselective entropy effects in the ligand exchange
reaction:

$$[M(H_2O)_6] + 2D/L-HisO \rightarrow [M(HisO)_2] + 6H_2O$$

are not likely to be great. With Cu(II) and Zn(II), on the other
hand, significant changes in coordination pattern on ligand exchange
are likely.

Various structures have been assigned to the Cu(II)-L-histidinate
complexes. The most likely structure for the bis complex is that shown
in Fig. 8, which has NNNO bonding in the coordination plane [50-52].

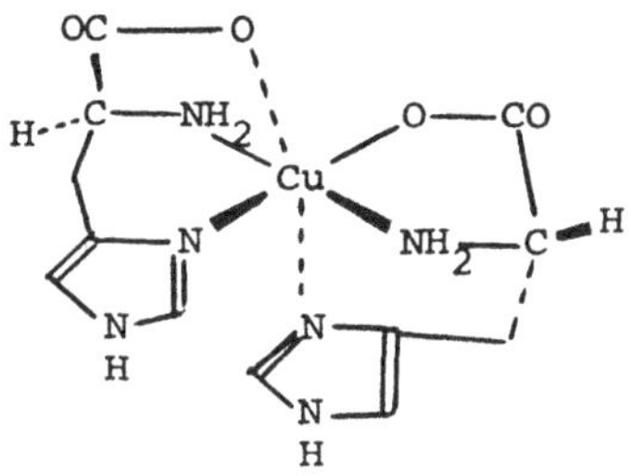

FIG. 8

It is interesting to note that the coordination scheme suggested in
Fig. 8 would lead to the positive stereoselectivity reflected in
enthalpy changes in favor of the optically active bis-copper complex,
as is found experimentally.

Although stereoselectivity is insignificant in the formation
constants of the Cu(II) bis complexes [Cu(D/L-HisO)$_2$], it can be
detected in the monoprotonated species, with the optically active
bis complex [Cu(L-HHisO)(L-HisO)]$^+$ being more stable than the meso
isomer [50]. The difference in stability is small (0.1 log units)
but significant, and has been reported also by Ritsma [53]. Similar
stereoselectivity is found in some ternary histidine and substituted
histidine complexes and will be discussed later.

Stereoselectivity has also been detected in a number of ternary
complexes of [Cu(HisO)]$^+$ with both substituted histidines and with
other amino acids. With N^3-benzyl-L-histidine and N^α,N^3-dibenzyl-
L-histidine stereoselectivity was small or absent in the unprotonated
ternary complexes but, with the monoprotonated species, formation
constants of the optically active species were greater by 0.12 ± 0.01
and 0.07 ± 0.01 log units respectively [54]; i.e., comparable to the
stereoselectivity found for histidine itself.

Studies of stereoselectivity in the ternary complexes [Cu(HisO)]$^+$
with other amino acids reveal two distinct trends, as demonstrated by
the results shown in Table 2 [55]. When the second amino acid is
potentially tridentate and the third donor center is a group that can
be protonated (e.g., an -NH$_3^+$ group) ternary complexes containing

TABLE 2

Stereoselectivity is the Formation Constants of
Ternary Complexes of Formula $[Cu(D/L-HisO)(L-AA)H_n]$
where $\Delta \log \beta = \log \beta_{CuLL} - \log \beta_{CuDL}$

AA	$\Delta \log \beta$ ($\pm$ 0.01)	
	n = 1	n = 0
L-ornithine	0.12	0
L-lysine	0.10	0
L-arginine	0.12	
L-aspartic acid	0.06	-0.08
L-phenylalanine	0	-0.20
L-tryptophan		-0.47
L-valine		0.06
L-leucine		0.03
L-threonine		0
L-serine		0
L-methionine		0
L-proline		0

ligands of the same chirality are more stable when protonated than
the meso isomers (e.g., ornithine, lysine, and arginine). Stereoselectivity in the neutral complexes is, however, insignificant.
With aspartic acid the opposite stereoselectivity to that with
ω-amino acids (and histidine or substituted histidines) is found--
the fully ionized meso complex is now the more stable isomer.

These results for the ω-amino acids may be explained by electrostatic interaction between the protonated $-NH_3^+$ side-chain and the
CO_2^- group of the histidinate ion which can take up an apical position
as shown in Fig. 9. If the structures of the complexes in solution
are similar to that found in the ternary complex [Cu(L-HisO)(L-ThrO)],
i.e., with the amino groups cis [30], then the $-NH_3^+$ and the histidine
CO_2^- groups are indeed on the same side of the Cu(II) ion for ligands

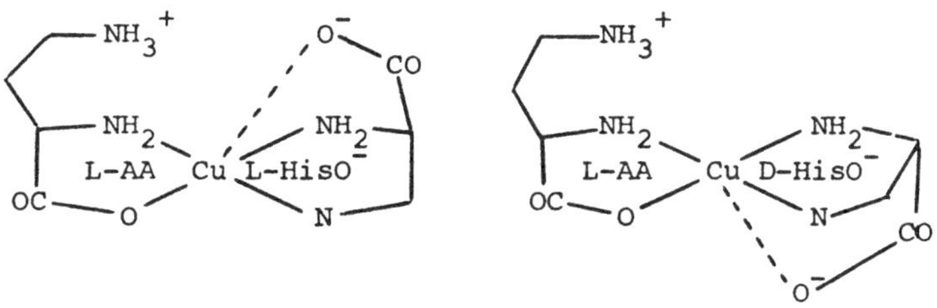

FIG. 9

of the same chirality. Stereoselectivity in the fully ionized
ternary complex with aspartic acid, this time favoring ligands of
opposite chirality, is therefore to be expected since the two CO_2^-
groups will prefer to align themselves on opposite sides of the
Cu(II) coordination plane.

When the second amino acid contains an aromatic ring (phenylalanine or tryptophan), the meso ternary complex is significantly more
stable than that containing ligands of the same chirality. With
valine, leucine, and typical nonbonding aliphatic groups the stereoselectivity was insignificant or tending to favor the optically
active ternary complex [55]. The reason for these forms of stereoselectivity is not immediately clear. It presumably results from
some aromatic ring-Cu(II) interactions or hydrophobic interactions.

An interesting application of stereoselectivity in ternary
histidine complexes is the resolution of racemic mixtures of a ligand
by selective crystallization of a ternary complex which forms stereoselectively. Using this principle DL-histidine has been resolved
via the ternary complex [Cu(L-HisO)(L-asparaginate)] to give an
optical purity of 98% [56].

Complexes of Ni(II) with histidine also show kinetic stereoselectivity. The rate of hydrolysis of L-histidine methyl ester was
found to be greater in the presence of [Ni(D-HisO)]$^+$ than [Ni(L-HisO)]$^+$
by about 40% [57]. This observation is particularly interesting
because it involves a 1:1, Ni(II)-histidine mixture when there is no
thermodynamic stereoselectivity in the formation of the [Ni(HisO)]$^+$
mono complexes. No such kinetic stereoselectivity was found with
Cu(II) [58]. Later work has shown that histidine methyl ester is

TABLE 3

Stereoselectivity in the Formation of Bis
Complexes of Some Substituted Histidines (A)
According to Reaction (4) (kJ mol^{-1})

	MeHis	DimeHis	BzHis
CuA$_2$	0	-2.34	1.09
CuHa$_2^+$	0.92	0	1.88
CoA$_2$	-2.88	-4.43	-1.17
NiA$_2$	-4.30	-6.77	-3.09
ZnA$_2$	-1.92	-5.0	0

hydrolyzed stereoselectively by a range of optically active Ni(II) catalysts, and that [Ni(HisO)]$^+$ will hydrolyse stereoselectively the methyl esters of lysine and tryptophan [59].

2.3.3. *Complexes of Substituted Histidines*

Stereoselectivity in complexes of substituted histidines is often larger than with histidine itself. Ritsma studied the complexes of N$^\alpha$-methyl-(MeHis), N$^\alpha$,N$^\alpha$-dimethyl-(DimeHis), and N$^\alpha$-benzylhistidine (BzHis) with Cu(II) [53], Co(II), Ni(II), and Zn(II) [60], and his results are summarized in Table 3. These results follow the same trend as those found with histidine, i.e., stereoselectivity with Co(II), Ni(II), and Zn(II) is negative (favoring the meso complex [M(L-AA)(D-AA)] whereas with Cu(II) it tends to be positive (favoring the optically active bis complex) in the monoprotonated species.

Because exactly the same trends are found in complexes of histidine, substituted histidines, and ternary complexes of these ligands, certain generalizations are possible [60]. Facially coordinating tridentate ligands in octahedral bis complexes can give efficient three-point attachment if the metal ion is not too big (e.g., Co(II), Ni(II), or to a lesser extent Zn(II) since this ion tends to be tetrahedral). Larger ions such as Cd(II) show much less

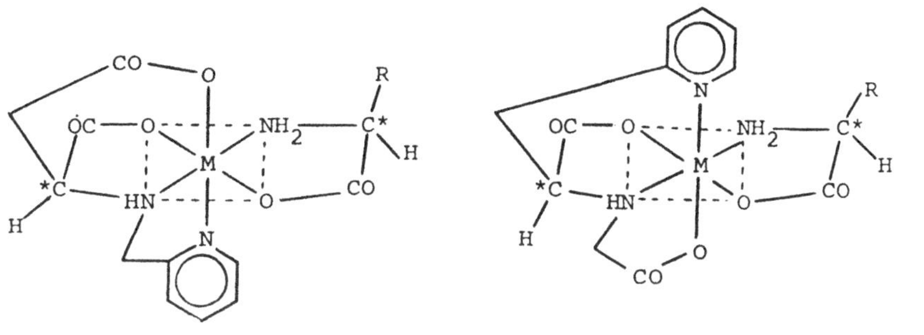

(a)   (b)

FIG. 10

significant stereoselective effects. Stereoselectivity tends to
increase as the degree of substitution increases and as the bulkiness of the substituents increases. Stereoselectivity in these cases
is, therefore, largely an enthalpy effect. With Cu(II), where the
planar coordination sites are the more important, the relative
stabilities of the cis and trans isomers will be important, affect-
ing the enthalpy changes, and differences in solvation will have a
significant effect on the entropy changes.

Histidine-like synthetic amino acids that have been studied
include β-(2-pyridyl)-α-alanine (Pyala) and β-(6-methyl-2-pyridyl)-
α-alanine (Mepyala) [61], N-(2-pyridylmethyl)-L-aspartic acid
(N-pyrasp) [62] and N-carboxymethyl-β-(2-pyridyl)-L-α-alanine
(Cmpyala) [63]. With Pyala and Mepyala stereoselective binding was
found with Zn(II), Co(II), and Ni(II), again the meso complex being
the more stable. With Cu(II) stereoselectivity was absent with
Pyala but significant in favor of the meso complex with Mepyala.

Complexes of N-pyrasp with Cu(II) and Ni(II) show stereoselective binding of optically active amino acids, the L isomer
being bound more firmly to give the optically active ternary complexes. This stereoselectivity has been rationalized in terms of
the preferred trans coordination of the amino nitrogen atoms to give
a complex with the geometry shown in Fig. 10a. On the other hand,
complexes of Cmpyala with L-amino acids would have the geometry
shown in Fig. 10b.

As a result of steric crowding, the side chain R would inter-
fere with the pyridyl group in 10b but not in 10a. Hence it would
be anticipated that, whereas N-pyrasp would bond L isomers preferentially, Cmpyala would prefer to react with D isomers to give a
meso ternary complex. This was, in fact, found experimentally to
be the case, giving elegant evidence for the steric origins of the
stereoselectivity found.

3. DIPEPTIDE COMPLEXES

3.1. Introduction

While complexes of dipeptides have been studied using several different experimental methods, stereoselectivity in the formation of these
complexes has been measured almost entirely by potentiometric methods.
This section will be concerned with complexes of H^+, Cu(II), and, to
a lesser extent, Ni(II) and Zn(II) with simple dipeptide, i.e., those
with side chains normally regarded as noncoordinating. To date, no
other work on stereoselectivity in diastereoisomeric dipeptide systems
involving other metal ions has been carried out.

It has been shown that dipeptides with only one optically
active center (e.g., glycyl-L-alanine) do not exhibit stereoselectivity in the formation of complexes with H^+ [64] or with some transition metal ions [65]. A dipeptide synthesized from optically pure
amino acids can exist as four isomers, divided into two pairs. The
LL and DD isomers are optically active or "pure" isomers, whereas
the LD and DL isomers are meso or "mixed" isomers. The members of
each pair are enantiomeric and the two pairs are diastereoisomeric.
It has been shown that the formation constants of the enantiomeric
dipeptides are indeed identical as expected [65, 66].

Before explanations of stereoselectivity can be undertaken,
the complexes present in the system must be identified clearly.
Incorrect assignment of formation constants has often been a source
of error in the past. A recent review [1] cites early work in which

the authors were apparently unaware of the ability of Cu(II) to
deprotonate the amide nitrogen of a dipeptide [67]. As a result,
the review relates values of the formation constants for the CuL
complex to an incorrect structure for the complex.

Incorrect assignment presents little problem in the proton
complexes. The structures of the complexes of Cu(II) with simple
dipeptides have been reasonably well elucidated, the major species
being $[CuL]^+$, $[Cu(L-1H)]$, and $[Cu(L-2H)]^-$. In addition a $[Cu(L-1H)$
$(L-2H)]^-$ complex is formed in solutions of a 1:1 metal-ligand ratio,
whereas a $[CuL(L-1H)]^-$ complex is formed in solutions with a 1:2
metal-ligand ratio. It has been established that in the $[CuL]^+$
complex Cu(II) coordinates to the amino nitrogen and amide oxygen.
The equivalent complexes of other transition metal ions are also
thought to coordinate similarly. The complex $[Cu(L-1H)]$ has a
basically planar structure and will bond through its amino and
deprotonated amide nitrogen atoms and the carboxyl oxygen (see
Fig. 3).

There is no doubt that the largest stereoselective effects
observed in these systems occur in the formation of the $[Cu(L-1H)]$
complex. There is some doubt however, as to whether any real stereoselectivity is observed in the formation of the $[CuL]^+$ and $[CuL(L-1H)]^-$
complexes. There appears to be no stereoselectivity in the stepwise
formation of the other complexes. The fact that there is some small
inconsistency in the results presented by different workers is of no
real importance for this review, which will discuss possible causes
for the large stereoselective effect found in the formation of the
$[Cu(L-1H)]$ complexes.

The Ni(II)-dipeptide systems are less well elucidated than
those with Cu(II), as a result of a tendency to precipitate at low
pH unless a high ligand-metal ratio is used. In practice it is found
that a number of possible models fit the data equally well and cannot
be elucidated by statistical or chemical methods [68]. However, the
important species $[NiL]^+$, $[NiL_2]$, and $[Ni(L-1H)]$ are unambiguously
defined. The Zn(II)-dipeptide system has even greater complications
as a result of polynuclear and hydroxy complex formation.

The observed stereoselectivity in these complexes with labile
metal ions is thermodynamic in origin. Kinetic stereoselectivity
has been reported in the formation of Co(III)-dipeptide complexes,
presumably resulting from steric factors [69].

3.2. Review of Results

Reported values for comparable formation constants are presented in
Tables 4 to 6. From the results for the formation constants of the
proton complexes (Table 4) it is clear that stereoselectivity in
these constants can be summarized for the carboxylic acid proton:

$$\log K \text{ (pure)} > \log K \text{ (meso)}$$

and for the amino proton:

$$\log K \text{ (meso)} > \log K \text{ (pure)}.$$

This means that the zwitterion form has a considerably greater pH
range of existence in the meso isomer than the pure. Some early
values have been omitted from Table 4 because they have been super-
seded by later and more accurate studies.

From the Cu(II) complexes, the [Cu(L-1H)] complex of the pure
dipeptide is almost always more stable than its meso analog. Stereoselectivity is considerably smaller or totally absent, in the [CuL]$^+$
complexes, the meso species tending to be the more stable. A similar
trend is found in the formation of Ni(II)-dipeptide complexes in the
two studies that have been reported [70, 71]. Both found the pure
[Ni(L-1H)] complex to be more stable than the meso isomer although,
relative to the Cu(II) system, the amide proton is ionized at a much
higher pH value. There is again some doubt as to the stereoselectivity in the [NiL]$^+$ complex. Nakon and Angelici observed that in
the Zn(II) systems the mixed [ZnL]$^+$ and [ZnL$_2$] complexes were more
stable than their pure isomers [66].

TABLE 4

Proton Complex Formation Constants of
Some Diastereomeric Dipeptides
at 25° and I = 0.10 M

	log K (CO_2^-)	log K (NH_2)	Ref.
L-Ala-L-Ala	3.30, 3.31	8.17, 8.15	66, 70
D-Ala-D-Ala	3.30	8.14	64
L-Ala-D-Ala	3.12	8.30	66
D-Ala-L-Ala	3.15	8.28	70
L-Ala-L-Phe	3.25	7.89	66
L-Ala-D-Phe	3.02	8.08	66
L-Lys-L-Ala	3.22	7.62	64
L-Lys-D-Ala	3.00	7.74	64
L-Lys-L-Lys	3.01	7.53	64
L-Lys-D-Lys	2.85	7.53	64
L-Leu-L-Leu	3.45, 3.46	7.91, 7.91	66, 65
L-Leu-D-Leu	3.05, 3.17	8.20, 8.28	66, 65
L-Leu-L-Tyr	3.23, 3.20	7.82, 7.73	66, 70
L-Leu-D-Tyr	2.96	8.30	66
D-Leu-L-Tyr	2.84	8.32	70
L-Val-L-Phe	3.19	7.89	71
L-Val-D-Phe	2.87	8.24	71
L-Phe-L-Val	3.40	7.41	71
L-Phe-D-Val	3.09	7.77	71
L-Leu-L-Ala	3.36	8.08	71
L-Leu-D-Ala	2.98	8.17	71
L-Pro-L-Ala	3.27	8.98	71
L-Pro-D-Ala	3.15	9.09	71
L-Pro-L-Phe	3.40	8.70	71
L-Pro-D-Phe	2.91	9.08	71
L-Met-L-Val	3.43	7.45	71
L-Met-D-Val	3.07	7.69	71
L-Met-L-Phe	3.14	7.29	71
L-Met-D-Phe	2.87	7.72	71
L-Tyr-L-Leu	3.40	7.41	71
L-Tyr-D-Leu	3.02	7.86	71

TABLE 5

Copper Complex Formation Constants of Some
Diastereoisomeric Dipeptides
at 25° and I = 0.10 M

	$\log \beta_{110}$	$\log \beta_{11-1}$	$\log K_{110}^{11-1}$	Ref.
L-Ala-L-Ala	5.54(1), 5.31(1)		3.72(1), 3.58(1)	66, 70
L-Ala-D-Ala	5.71(2)	1.75	3.96(2)	66
D-Ala-L-Ala	5.60(1)	1.56	4.04(1)	70
L-Ala-L-Phe	5.20(2)	1.76	3.44(2)	66
L-Ala-D-Phe	5.42(2)	1.49	3.93(2)	66
L-Leu-L-Leu	5.21(1), 5.24(2)		3.88(1), 3.86(2)	66, 65
L-Leu-D-Leu	5.48(2), 5.20(3)		4.88(2), 4.57(3)	66, 65
L-Leu-L-Tyr	5.15(2), 5.19(1)		3.38(2), 3.27(1)	66, 70
D-Leu-L-Tyr	5.40(2), 5.34(1)		4.09(2), 4.08(1)	66, 70
L-Val-L-Phe	5.35(5)	1.845(2)	3.50	71
L-Val-D-Phe	5.31(1)	1.395(1)	3.91	71
L-Phe-L-Val	4.65(3)	1.155(1)	3.45	71
L-Phe-D-Val	4.86(3)	0.939(2)	3.92	71
L-Leu-L-Ala	5.62(2)	1.845(1)	3.77	71
L-Leu-D-Ala	5.56(2)	1.384(1)	4.18	71
L-Pro-L-Ala	6.60(1)	2.936(1)	3.66	71
L-Pro-D-Ala	6.48(1)	2.654(1)	3.83	71
L-Pro-L-Phe	6.53(1)	3.271(2)	3.26	71
L-Pro-D-Phe	6.23(5)	2.551(2)	3.68	71
L-Met-L-Val	4.96(1)	1.161(1)	3.22	71
L-Met-D-Val	5.01(1)	0.788(1)	3.80	71
L-Met-L-Phe	4.76(4)	1.758(2)	3.00	71
L-Met-D-Phe	4.93(2)	1.294(1)	3.64	71
L-Tyr-L-Leu	4.93(2)	1.065(5)	3.86	71
L-Tyr-D-Leu	4.88(1)	0.471(5)	4.41	71

Some $\log K_{110}^{11-1}$ values for other Cu(II)-dipeptides [65, 71]

Gly-Gly	Gly-Val	Gly-Phe	Gly-Met	Phe-Gly	Pro-Gly	Leu-Gly
4.09	4.64	3.88	4.09	3.75	4.01	3.90

TABLE 6

Nickel and Zinc Complex Formation Constants
of Some Diastereoisomeric Dipeptides
at 25° and I = 0.10 M

	$\log \beta_{110}$	$\log \beta_{120}$	$\log K_{110}^{11-1}$
Ni(II) [71]			
L-Ala-L-Ala [70]	4.14(1)	7.02(3)	8.67(3)
D-Ala-L-Ala [70]	3.90(1)	6.92(3)	9.06(3)
L-Leu-L-Tyr [70]	3.28(2)	-	8.06(3)
D-Leu-L-Tyr [70]	3.44(2)	6.46(4)	8.89(3)
L-Val-L-Phe	3.19(1)	5.39(3)	8.50(1)
L-Val-D-Phe	3.24(1)	6.32(1)	9.30(2)
L-Phe-L-Val	2.77(2)	5.21(4)	8.65(4)
L-Phe-D-Val	3.08(1)	6.08(1)	9.28(1)
L-Leu-L-Ala	3.36(1)	5.97(1)	8.92(1)
L-Leu-D-Ala	3.34(1)	6.15(1)	9.20(2)
L-Met-L-Val	3.21(4)	6.31(2)	8.88(6)
L-Met-D-Val	3.23(3)	6.26(2)	9.15(5)
L-Met-L-Phe	3.13(1)	5.78(1)	8.08(1)
L-Met-D-Phe	3.51(2)	6.44(4)	8.58(3)
L-Pro-L-Ala	4.41(1)	7.88(1)	8.33(1)
L-Pro-D-Ala	4.44(1)	8.67(1)	8.34(1)
L-Pro-L-Phe	4.25(1)	-	7.76(1)
L-Pro-D-Phe	4.05(1)	8.65(1)	8.53(2)
L-Tyr-L-Leu	3.14(1)	-	8.73(1)
L-Tyr-D-Leu	3.53(1)	-	9.15(1)
Zn(II) [66]			
L-Ala-L-Ala	3.73(2)	6.88(4)	
L-Ala-D-Ala	3.87(1)	7.04(3)	
L-Ala-L-Phe	3.38(1)	6.20(4)	
L-Ala-D-Phe	3.61(2)	6.55(4)	
L-Leu-L-Tyr	3.36(1)	6.32(2)	
D-Leu-L-Tyr	3.89(2)	7.09(4)	

3.3. Discussion of Results

3.3.1. Complexes of H^+

Ellenbogen interpreted stereoselectivity in the proton complex formation constants (Table 4) as resulting from differences in the ease
of folding the two diastereoisomers [64]. It can be seen from molecular models that the meso isomer will fold more easily than the pure
isomer. In the folded conformation of the zwitterion the oppositely
charged NH_3^+ and CO_2^- groups are stabilized by electrostatic interaction. Hence the meso isomer would be expected to be stabilized
as the zwitterion more than the pure isomer, giving the stereoselectivity found.

Nakon and Angelici observed that this stereoselectivity increased as the size of the side chain increased [66]. They accounted
for the observed stereoselectivity by assuming that the dipeptide
remained in the folded β conformation (see Fig. 11) in acidic, neutral, and basic solutions. This assumption was suggested and supported by the nmr work of Lemieux and Barton [72] and was consistent
with dielectric measurements [73]. Assuming the conformation shown
in Fig. 11, it can be seen that protonation of the carboxyl group is
electrostatically less stable in the meso isomer than in the pure
isomer due to the close proximity of the positively charged NH_3^+
group. For a similar reason deprotonation of the NH_3^+ group is
delayed in the meso isomer. The authors suggested the concept of
hydrophobic bonding to account for the increase in size of the
stereoselectivity with increasing size of the side chain. It was
suggested that the larger side chains promote hydrophobic bonding

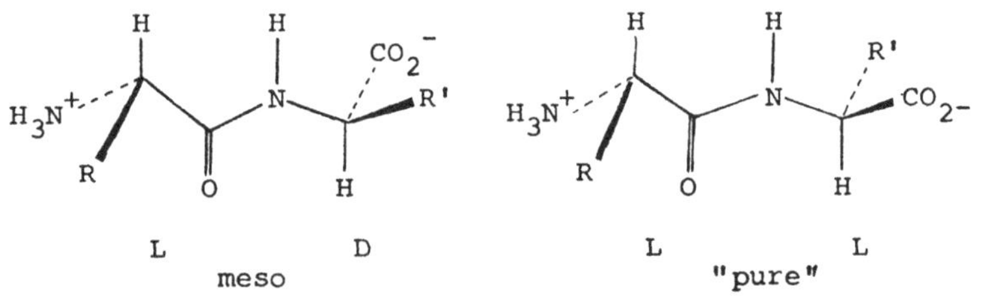

FIG. 11

between the two alkyl groups and the amide group, and that this
"attraction" may compress the molecule in the meso isomer. This
would have the effect of bringing the charged groups closer to each
other in the meso isomer, thus enlarging the stereoselectivity.

Kaneda and Martell observed similar stereoselectivity in the
two diastereoisomeric dipeptide systems they studied [70]. They
also explained the stereoselectivity as a result of a conformational
analysis carried out with the aid of molecular models. If the con-
formationally preferred form had a cis configuration, with both side
chains on the same side of the molecule, it was said to have a more
extensive hydrophobic region than the trans configuration with side
chains on opposite sides of the molecule. The resulting organiza-
tional effect on the surrounding water molecules, restricting trans-
lational and rotational freedom, was said to represent a lowering
of entropy. A change in configuration from trans to cis therefore
involved a decrease in entropy. Hence the trans form would be
expected to be the more stable. The configurations of the preferred
forms, as determined by conformational analysis of the proton complexes, was found to be:

	H_2L^+	HL	L^-
pure	trans	cis	trans
meso	cis	trans	cis

If it is assumed that enthalpies of protonation are not stereoselective, the meso zwitterion will be the more stable, making the
above explanation consistent with the observed protonation constants.
In practice, as the authors state, the explanation cannot be as
simple as this because the actual configurations will not be entirely
cis or trans and enthalpies of protonation will not be entirely independent of changes in conformation. The concept of hydrophobic
bonding is discussed further in Sec. 3.4.

3.3.2. Complexes of Cu(II)

Various explanations have been given for stereoselectivity in the formation of the [Cu(L-1H)] complex. The metal ion and the three donor atoms are known to be virtually coplanar in the solid state [74], and it can be assumed that this will also be true in solution as shown in Fig. 3. Two axial water molecules will be relatively loosely bound in a tetragonally distorted position. In the optically pure complex both side chains will be situated on the same side of the coordination plane, in the meso complex they will be on opposite sides. The pure complex is known to be more stable than the meso isomer--a result not expected on the basis of steric interference between the side chains. One explanation assumed that the axial water-side chain interactions were unfavorable, making the meso complex, with two such interactions, less favorable than the pure complex with only one interaction [65]. This explanation would make stereoselectivity the result of differences in destabilizing effects. Deprotonation of [Cu(L-1H)] to give [Cu(L-2H)]$^-$ involves hydrolysis of a coordinated water molecule. This reaction would not, therefore, be expected to be stereoselective.

Nakon and Angelici explained the stereoselectivity in [Cu(L-1H)] as resulting from an enhanced stability of the pure isomer relative to the meso isomer, as a direct result of hydrophobic bonding [66]. They suggested that, in the pure complex in which the two side chains are on the same side of the molecule, hydrophobic bonding may occur between the two alkyl groups and the amide linkage forming an "internal micelle." This would modify the solvent structure, giving a more favorable water-complex interface. In the meso complex the side chains are on opposite sides. Hence such a large effect cannot occur. They also detected stereoselectivity in the formation of the [CuL]$^+$ complexes. The explanation given for this was similar to that given for the proton complexes.

The explanation given by Kaneda and Martell for stereoselectivity in the Cu(II) complexes is similar to that given for the proton complexes [70]. The conformational analysis was used to

decide whether the constituents on the chelate rings were axial or
equatorial, then the entropy argument was applied. This argument
assumes a deviation from the planar chelate rings normally expected
in such complexes so that the rings are twisted sufficiently to
form δ or λ chelate rings. The amide linkage is thought to retain
its planarity as a result of resonance stabilization, making it
difficult to accept that there is sufficient deviation from planarity
in solution to permit δ or λ isomers. Hence this explanation of the
observed stereoselectivity may not be entirely complete.

3.4. Hydrophobic Interactions

Hydrophobic interactions cannot be classified as bonds but they are
important, not well understood, and often misused. Although the
existence of the interactions is not in doubt [75], their relevance
to the complexes of organic molecules in aqueous solution is not
clear. In aqueous solution hydrophobic portions of molecules will
tend to aggregate together to minimize unfavorable energy effects
at the organic-water interface. The effect, therefore, does not
arise out of any attraction of the chains for each other, but out
of the strong attractive forces that exist between water molecules.

Hydrophobic interactions may be used to explain two opposite
effects. For example Kaneda and Martell state that a relatively
more extensive hydrophobic region results in an entropy loss and
in a relative lowering of stability [70]. Nakon and Angelici, on
the other hand, suggest that by creating an internal micelle, with
a large hydrophobic region, it is possible to decrease an energeti-
cally unfavorable solvent-complex interface, leading to the formation of a more stable complex [66]. The latter suggestion is probably the one to be preferred as it is not the size of the hydrophobic
region that is important so much as the area of solvent that has
contact with the hydrophobic portions of the complex. This area is
less when the hydrophobic chains lie next to one another. Hence,
from the former argument, there should be a relative entropy gain

when the larger hydrophobic region is formed, not an entropy loss.
However, the incorporation of the amide group, with its large dipole
moment, in a hydrophobic region may not itself be totally valid.

3.5. Some Comments and Suggestions

Observed stereoselectivity can only be satisfactorily discussed in
terms of the conformations of the dipeptide molecule in all its
complexes. Much work has been devoted to the study of polypeptide
as well as dipeptide conformations [76, 77]. The work of Beecham
et al., who showed that the two charged groups of the zwitterion
were further away in the pure dipeptide than the meso analog [73],
provides a satisfactory explanation for stereoselectivity in proton
complexes. However it is probably true to say that more than one
conformation will contribute to the final population of structures
in solution [76].

   Several different effects must be considered in relation to
stereoselectivity.

   1. Hydrophobic effects, which have been discussed already.
   2. Electrostatic effects between charged groups. Charges
      will tend to be shielded in solutions of high ionic
      strength. Hence these effects will be very concentration
      dependent and will be particularly important with dipep-
      tides containing reactive side chains.
   3. Steric effects between the side chains themselves. These
      effects will, in general, be destabilizing effects.
   4. Effects on the surrounding solvent sphere, caused by the
      positioning of the hydrophilic and hydrophobic parts of
      the total complex.
   5. Dipole-induced dipole interactions, which may be important
      in complexes containing aromatic moieties.

   All of these effects may be discussed in a qualitative manner
but it is very difficult to predict how they will finally produce a
structure for a complex ion in solution.

All the evidence suggests that the conformations of the mono-proton complex in its zwitterion form have the two charged groups on the same side of the molecule in the meso isomer and on opposite sides in the pure isomer. However, it is difficult to see why the pure isomer does not also have a majority of its conformers with both charged groups on the same side also. The reason is probably because, in this conformation, the two C_α-C_β bonds of the side chains would be eclipsed and therefore sterically unfavorable. The stereo-selectivity found suggests, therefore, that the electrostatic inter-action between the charged groups of the zwitterion may not be as strong or as important as has sometimes been suggested. Whatever the specific effects which cause stereoselectivity in the proton complex formation constants, they cannot all be present in complexes of $[CuL]^+$ because, in this complex, the dipeptide must exist in a different conformation. As can be seen from Table 5, we found sig-nificant stereoselectivity (> 3σ) in the formation of this complex in only one Cu(II)-dipeptide system, that with methionylphenylalanine. We consider this to be due to an unusual factor of this particular system [71]. We suggest that stereoselectivity in $[CuL]^+$ complexes of *simple* dipeptides is not a general phenomenon.

An understanding of the stereoselectivity in the $[Cu(L-1H)]$ complexes is helped by considering the acidity of the $[CuL]^+$ complex of glycylglycine and the corresponding glycine containing dipeptides, and comparing these values with the acidities of the diastereoiso-meric $[CuL]^+$ complexes. Some appropriate values for log K_{110}^{11-1} are included in Table 5. From these it can be seen that the cause of the stereoselectivity must be a stabilizing effect [65]. The pure $[CuL]^+$ complexes are always more acidic than the glycine-containing analogs whereas the meso complexes are of comparable or possibly slightly larger acidity. It is also interesting to note that the most acidic of the pure $[CuL]^+$ complexes are those containing a potentially coordinating side chain, such as a thioether group or phenyl ring. The thioether group can coordinate weakly to Cu(II) along the z axis, but it only appears to do so if the thioether group is in the N-terminal residue of the dipeptide. It is possible that disruption of the solvent sphere around the ionic carboxylate group could be the reason for the absence of comparable gains in stability when the thioether group is in the O-terminal residue [71]. This suggestion of a thioether-Cu(II) bond is of special interest as a result of the recent discovery of such an interaction by the X-ray diffraction study of the blue copper protein plastocyanin [78].

Aromatic groups appear to confer extra stability to certain complexes by a weak association over the ionic center of the complex, giving a dipole-induced dipole interaction. Such interaction has recently been confirmed by visible and epr spectroscopy [79]. These interactions have been found in Cu(II)-dipeptide systems but not in Cu(II)-amino acid systems [80]. Hence the stereoselectivity found in complexes of methionylphenylalanine would be the result of a number of additional contributing factors.

An interesting example of stereoselectivity is found in the binuclear complex of Cu(II), Ag(I), and methionylmethionine. A discrete molecule $[CuAg(L-Met-L-MetO)]^+$ forms with the pure isomer while the meso isomer only forms an insoluble polymer [44]. The two sulfur atoms are able to bond effectively to Ag(I) in the pure complex because both side chains are on the same side of the Cu(II) coordination plane. Effective bonding to Ag(I) is impossible with the meso isomer.

Stereoselectivity in the Ni(II)-diastereoisomeric dipeptide systems is probably the result of similar factors to those discussed for Cu(II). However, it should be remembered that the coordination sphere of Ni(II) will approximate more nearly to a regular octahedron, drawing the axial water molecules much closer to the central metal ion than is found with Cu(II).

REFERENCES

1. K. Bernauer, *Topics in Current Chemistry, 65,* 1 (1976).

2. A. T. Advani, H. M. N. H. Irving, and L. D. Pettit, *J. Chem. Soc., A,* 2649 (1970); R. D. Gillard, H. M. N. H. Irving, and L. D. Pettit, *J. Chem. Soc., A,* 673 (1968).

3. G. R. Brubaker, *J. Chem. Ed., 51,* 608 (1974).

4. J. L. Meyer and J. E. Bauman, *J. Chem. Eng. Data, 15,* 404 (1970).

5. W. A. Franks and D. Van der Helm, *Acta Crystallogr., B27,* 1299 (1970); D. Van der Helm and C. E. Tatsch, ibid, *B28,* 2307 (1972).

6. G. G. Aleksandrov, Yu. T. Struchkov, A. A. Kurganov, S. V. Rogozhin, and V. A. Davankov, *Chem. Comm.,* 1328 (1972).

7. E. J. Corey and J. C. Bailar, *J. Amer. Chem. Soc., 81,* 2620 (1959).

8. J. R. Gollogly and C. L. Hawkins, *Inorg. Chem., 8,* 1168 (1969); *9,* 576 (1970).

9. A. T. Advani, D. S. Barnes and L. D. Pettit, *J. Chem. Soc., A,* 2691 (1970).

10. L. D. Pettit and J. L. M. Swash, *J. Chem. Soc. Dalton,* 697 (1977).

11. J. E. Sarneski and C. N. Reilley, *Inorg. Chem., 13,* 977 (1974).

12. J. H. Dunlop, D. F. Evans, R. D. Gillard, and G. Wilkinson, *J. Chem. Soc., A,* 1260 (1966).

13. L. D. Pettit and J. L. M. Swash, *J. Chem. Soc. Dalton,* 286 (1978).

14. K. Bernauer, Metal Ions in Biological Systems, Vol. 1, (H. Sigel, ed.), Dekker, New York, 1974.

15. S. V. Rogozhin and J. A. Davankov, *Chem. Comm.,* 490 (1971); *J. Chromatogr., 60,* 280 (1971).

16. R. V. Snyder, R. J. Angelici, and R. E. Meck, *J. Amer. Chem. Soc., 94,* 2660 (1972).

17. K. Bernauer, M. F. Jeanneret, and D. Vonderschmidt, *Helv. Chim. Acta, 54,* 297 (1971).

18. H. Takenaka and M. Shibata, *Bull. Chem. Soc. Jap., 49,* 2133 (1976).

19. G. Colomb and K. Bernauer, *Helv. Chim. Acta, 60,* 459 (1977).

20. K. M. Wellman, T. G. Mecca, W. Mungall, and C. R. Hare, *J. Amer. Chem. Soc., 89,* 3646 (1967).

21. R. D. Gillard, H. M. N. H. Irving, R. Parkins, N. C. Payne, and L. D. Pettit, *J. Chem. Soc., A,* 1159 (1966).

22. V. Simeon and O. A. Weber, *Croat. Chim. Acta, 38,* 161 (1966); *Biochem. Biophys. Acta, 244,* 94 (1971).

23. V. A. Davankov and P. R. Mitchell, *J. Chem. Soc. Dalton,* 1012 (1972).

24. G. Brookes and L. D. Pettit, *J. Chem. Soc. Dalton,* 1918 (1977).

25. V. A. Davankov, S. V. Rogozhin, Yu. T. Struchkov, G. G. Aleksandrov, and A. A. Kurganov, *J. Inorg. Nucl. Chem., 38,* 631 (1976).

26. V. A. Davankov, S. V. Rogozhin, A. A. Kurganov, and L. Ya Zhuchkova, *J. Inorg. Nucl. Chem., 37,* 369 (1975).

27. R. V. Snyder and R. J. Angelici, *J. Inorg. Nucl. Chem., 35,* 523 (1973).

28. B. E. Leach and R. J. Angelici, *J. Amer. Chem. Soc., 91,* 6296 (1969).

29. J. H. Ritsma, *Rec. Trav. Chim., 94,* 174 (1975).

30. H. C. Freeman, J. M. Guss, M. J. Healy, R. P. Martin, C. E. Nockholds, and B. Sarkar, *Chem. Comm.,* 225 (1969).

31. H. C. Freeman and R. P. Martin, *J. Biol. Chem., 244,* 4823 (1969).

32. B. Sarkar, M. Bersohn, Y. Wingfield and T. C. Chiang, *Can. J. Biochem., 46,* 595 (1968).

33. L. D. Pettit and J. L. M. Swash, *J. Chem. Soc. Dalton,* 2416 (1976).

34. D. Van der Helm and W. A. Franks, *Acta Crystallogr., B25,* 451 (1969); D. Van der Helm and M. B. Hossain, ibid., *B25,* 457 (1969).

35. J. H. Ritsma, G. A. Wiegers, and F. Jellinek, *Rec. Trav. Chim., 84,* 1577 (1965).

36. P. J. Morris and R. B. Martin, *J. Inorg. Nucl. Chem., 32,* 2891 (1970).

37. K. Kawasaki, J. Yoshii, and M. Shibata, *Bull. Chem. Soc. Jap., 43,* 819 (1970); T. Matsuda, and M. Shibata, ibid., *46,* 3104 (1973).

38. O. Yamauchi, T. Sakurai, and A. Nakahara, *Bull. Chem. Soc. Jap., 50,* 1176 (1977).

39. G. R. Lenz and A. E. Martell, *Biochemistry, 3,* 745 (1965).

40. M. V. Veidis and G. J. Palenik, *Chem. Comm.,* 1277 (1969).

41. D. B. McCormick, H. Sigel, and L. D. Wright, *Biochem. Biophys. Acta, 184,* 318 (1969).

42. J. L. M. Swash and L. D. Pettit, *Inorg. Chim. Acta, 19,* 19 (1976).

43. C. A. McAuliffe, J. V. Quagliano, and L. M. Vallarino, *Inorg. Chem., 5,* 1996 (1966).

44. L. D. Pettit and K. F. Siddiqui, Proc. XIX Internat. Conf. Coord. Chem., Prague, 1978, p. 98.

45. O. A. Weber, *Int. J. Prot. Res., 3,* 255 (1971).

46. J. H. Ritsma and F. Jellinek, *Rec. Trav. Chim., 91,* 923 (1972).

47. C. C. McDonald and W. D. Phillips, *J. Amer. Chem. Soc., 85,* 3736 (1963).

48. J. H. Ritsma, J. C. Van de Grampel, and F. Jellinek, *Rec. Trav. Chim., 88,* 411 (1969).

49. D. S. Barnes and L. D. Pettit, *J. Inorg. Nucl. Chem., 33,* 2177 (1971).

50. L. D. Pettit and J. L. M. Swash, *J. Chem. Soc. Dalton,* 588 (1976).

51. H. Sigel and D. B. McCormick, *J. Amer. Chem. Soc., 93,* 2041 (1971).

52. G. Rotilio and L. Calabrese, *Arch. Biochem. Biophys., 143,* 218 (1971).

53. J. H. Ritsma, *Rec. Trav. Chim., 94,* 210 (1975).

54. G. Brookes and L. D. Pettit, *J. Chem. Soc. Dalton,* 1224 (1976).

55. G. Brookes and L. D. Pettit, *J. Chem. Soc. Dalton,* 1918 (1977).

56. T. Sakurai, O. Yamauchi and A. Nakahara, *Chem. Comm.,* 718 (1977).

57. J. E. Hix and M. M. Jones, *J. Amer. Chem. Soc., 90,* 1723 (1968).

58. R. W. Hay and P. J. Morris, *Chem. Comm.,* 18 (1969).

59. J. R. Blackburn and M. M. Jones, *J. Inorg. Nucl. Chem., 35,* 1597, 1605, 2421 (1973).

60. J. H. Ritsma, *J. Inorg. Nucl. Chem., 38,* 907 (1976).

61. P. R. Rechani, R. Nakon, and R. J. Angelici, *Bioinorganic Chem. 5,* 329 (1976).

62. R. Nakon, P. R. Rechani, and R. J. Angelici, *Inorg. Chem., 12,* 2431 (1973).

63. S. A. Bedell, P. R. Rechani, R. J. Angelici, and R. Nakon, *Inorg. Chem., 16,* 972 (1977).

64. E. Ellenbogen, *J. Amer. Chem. Soc., 74,* 5198 (1952); *78,* 369 (1956).

65. G. Brookes and L. D. Pettit, *J. Chem. Soc. Dalton,* 2303 (1975).

66. R. Nakon and R. J. Angelici, *J. Amer. Chem. Soc.*, *96*, 4178 (1974).

67. F. Karczinsky and G. Kupriszewski, *Rocz. Chem.*, *41*, 1019 (1967).

68. G. Brookes and L. D. Pettit, *J. Chem. Soc. Dalton*, 2106 (1975).

69. P. J. Morris and R. B. Martin, *Inorg. Chem.*, *10*, 964 (1971).

70. A. Kaneda and A. E. Martell, *J. Amer. Chem. Soc.*, *99*, 1586 (1977).

71. R. J. W. Hefford and L. D. Pettit, to be published.

72. R. U. Lemieux and M. A. Barton, *Can. J. Chem.*, *49*, 767 (1971).

73. J. Beecham, V. T. Ivanov, G. W. Kenner, and R. C. Sheppard, *Chem. Comm.*, 386 (1965).

74. H. C. Freeman, *Adv. Prot. Chem.*, *22*, 336 (1967).

75. C. Tanford, The Hydrophobic Effect, Wiley-Interscience, New York, 1973.

76. P. Gupta-Byaya, *Biopolymers*, *14*, 1143 (1975).

77. C. B. Anfinsen and H. A. Scheraga, *Adv. Prot. Chem.*, *29*, 205 (1975); B. Pullman and A. Pullman, ibid., 348 (1974).

78. P. M. Coleman, H. C. Freeman, J. M. Guss, M. Murata, V. A. Norris, J. A. M. Ramshaw, and M. P. Venkatappa, *Nature*, *272*, 319 (1978).

79. L. Sportelli, H. Neubacher and W. Lohman, *Z. Naturforsch*, *32C*, 643 (1977).

80. L. Sportelli, H. Neubacher, and W. Lohman, *Rad. Environ. Biophys.*, *13*, 305 (1976).

Chapter 7

PROTONATION AND COMPLEXATION OF MACROMOLECULAR POLYPEPTIDES: CORTICOTROPIN FRAGMENTS AND BASIC TRYPSIN INHIBITOR (KUNITZ BASE)

Kálmán Burger
Institute of Inorganic and Analytical Chemistry
L. Eötvös University
Budapest, Hungary

1. INTRODUCTION

The coordination chemical study of biologically active macromolecules, such as polypeptides, proteins, etc., has equally great importance from the biochemical, pharmacological, and pharmacotechnological points of view. Changes in the protonation of these molecules or the coordination of metal ions by their donor groups may influence their biological activity by changing the structure and charge of the species in solutions, resulting in changes in their transport rate and possibly mechanism in vivo, and also in the affinity of these molecules toward receptors in the living cell.

Metal complex formation might hinder the enzymatic decomposition of some peptides. This effect is utilized in the employment of metal ions in protein purification processes. The retarding effect of some polypeptide-containing medicines (corticotropin, insulin) can be ensured also by the use of their zinc complexes.

The investigation of the polypeptide complexes of such metals as zinc, iron, cobalt, etc., that can be incorporated without undesirable physiological effects is particularly important.

Metal ions can be used also as structural probes by which the accessibility and reactivity of potential donor groups on proteins can be tested. The coordination of silver ion might be, e.g., the indicator for the accessibility of sulfur-containing donor groups.

Changes in the protonation of macromolecular polypeptides might result in the formation or cleavage of intramolecular hydrogen bridges which stabilize the conformation of the molecule. Thus the shift in protonation-deprotonation equilibria might be connected with conformational transformation of the peptide resulting also in changes of its biological activity. A knowledge of the protonation equilibria is also of fundamental importance in the quantitative study of metal complex formation.

The great number of donor groups and the sterical effect of the ordered conformation (helix, etc.) of the macromolecules makes the exact evaluation of the equilibrium data of this type of system

difficult. That is why the systematic study of the coordination
chemical equilibria of macromolecular polypeptides started only 10
to 15 years ago [1].

The use of precision potentiometric equilibrium measurements
in the broadest possible concentration and pH ranges and the application of computer analysis for the evaluation of the experimental
data make the determination of the composition of the species (metal-
peptide or proton-peptide ratios) in the system possible. The assignment of the functional groups to the successive equilibrium steps on
the basis of only such investigations remains uncertain, and this
makes essential the combination of different experimental methods
in the study of the complex formation of macromolecules.

This chapter presents the systematic equilibrium study of
corticotropin (ACTH) fragments and basic bovine pancreatic trypsin
inhibitor (BPTI), two biologically important systems.

1.1. Corticotropin (ACTH)

The natural human adrenocorticotrophic hormone α_H-corticotropin
(ACTH) is of vital importance in the human organism [2]. It is a
polypeptide consisting of 39 amino acids. Its amino acid sequence
is known (see Fig. 1) [3, 4] but conflicting views can be found
with regard to its conformation and its secondary and tertiary
structures.

Several research groups first investigated the conformation
of the natural hormone and then, following the first synthetic preparation of ACTH [5], that of various synthetic ACTH fragments.

From the results of optical rotatory dispersion (ORD) [6] and
circular dichroism (CD) [7] studies in aqueous solutions and of
hydrogen-deuterium exchange experiments [8, 9], it was concluded
that in aqueous solutions the polypeptide molecule exhibits an
entirely disordered form. However, recent CD studies [10, 11] in
agreement with earlier considerations [12] suggested the ordering

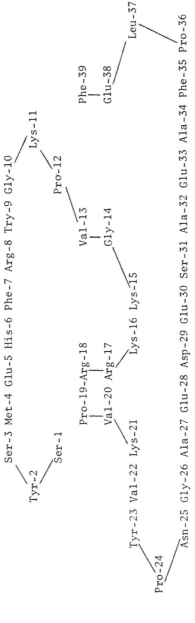

FIG. 1. Amino acid sequence of corticotropin.

of the molecule even in aqueous solution and demonstrated the presence of an α-helix structure in trifluoroethanol.

The pH dependence of the CD curves of ACTH in aqueous solution
[11] showed that ordering, probably stabilized by the side chains of
the polypeptide, occurs on the increase of the pH.

Since already smaller fragments of corticotropin, e.g., the
N-terminal 24 or 32 amino acid-containing fragment, also have a
biological activity equal to that of the natural hormone, investigation of these fragments became essential.

The corticotropin polypeptide fragment, $ACTH_{1-32}$, discussed
in the following, consists of 32 amino acids and contains 15 different
amino acids. Its complex-forming functional groups are the peptide
bonds, five carboxyl, five primary amino and three guanidino groups,
two phenolic hydroxy and three alcoholic hydroxy groups, one imidazole nitrogen, one indole nitrogen, and one thioether sulfur atom.

The protonation-deprotonation [13] and the complex formation
equilibria with zinc(II) [14] and silver ions of $ACTH_{1-32}$ [15] were
studied. The details of these investigations are presented in
Secs. 2.1, 2.2, 3.1, and 3.2.

1.2. Basic Bovine Pancreatic Trypsin Inhibitor (BPTI)

Basic pancreatic trypsin inhibitor, i.e., kallikrein inactivator or
Kunitz base, was studied far more intensively. This polypeptide
consists of 58 amino acids and contains 18 different amino acids.
Its amino acid sequence has been known for some time [16-18] (Fig. 2)
and its atomic structure has been determined by X-ray diffraction
studies [19-21] (Fig. 3).

Like ACTH, BPTI can be considered a polyfunctional ligand.
Its complex-forming functional groups are the six guanidino, four
phenolic hydroxy, five primary amino, and five carboxyl groups, one
thioether sulfur atom and three disulfide bridges.

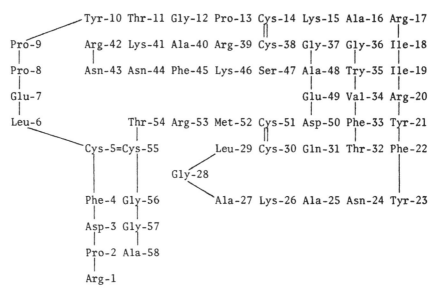

Tyr-10 Thr-11 Gly-12 Pro-13 Cys-14 Lys-15 Ala-16 Arg-17

Pro-9   Arg-42 Lys-41 Ala-40 Arg-39 Cys-38 Gly-37 Gly-36 Ile-18

Pro-8   Asn-43 Asn-44 Phe-45 Lys-46 Ser-47 Ala-48 Try-35 Ile-19

Glu-7   Glu-49 Val-34 Arg-20

Leu-6   Thr-54 Arg-53 Met-52 Cys-51 Asp-50 Phe-33 Tyr-21

         Cys-5=Cys-55   Leu-29 Cys-30 Gln-31 Thr-32 Phe-22

                         Gly-28

                               Ala-27 Lys-26 Ala-25 Asn-24 Tyr-23

         Phe-4 Gly-56

         Asp-3 Gly-57

         Pro-2 Ala-58

         Arg-1

FIG. 2. Amino acid sequence of BPTI.

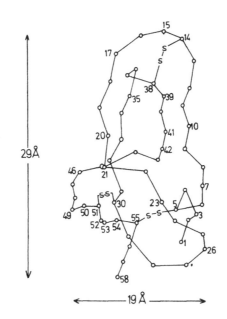

FIG. 3. Atomic structure of BPTI.

BPTI is reasonably resistant to chemical environments not often tolerated by many other polypeptides. Its reaction with trypsin and kallikrein was studied by several research groups. Even the equilibrium constants of the association reaction of BPTI with trypsin and kallikrein [22-24] were determined. Later on the structure of this complex was evaluated by means of X-ray diffraction [25, 26]. The protonation [27] and silver complex formation equilibria [28] of BPTI were studied in detail. The results of these investigations are presented in Secs. 2.3 and 3.3.

The interactions of calcium and iron ions with BPTI were also investigated [29]. It was shown that calcium ions are not bound by this peptide in aqueous solutions of pH 6 to 10, containing a calcium content about 10 to 50 times higher than the BPTI concentration.

The iron complex formation of BPTI was studied by Mössbauer spectroscopy [29]. These investigations have indicated the stabilization of the iron(II) oxidation state in the complex.

2. PROTONATION EQUILIBRIA

Intramolecular hydrogen bridges stabilize the secondary and tertiary structure of macromolecular polypeptides. Because hydrogen bond formation changes the dissociation constants of the protons involved in this bond and of protons affected in other ways by the formation of the H bridges, the protonation constants can be considered to give also structural information on the systems [13].

The corresponding equilibrium constant can be determined by high-precision pH metric measurements in accordance with the principles of classical equilibrium chemistry [30, 31]. The solutions to be examined have to be adjusted to a constant ionic strength high enough to allow for the neglect of the polypeptide content contribution. The peptide is protonated with a known quantity of acid. The emf change caused by the pH increase of the solution during titration with a standard NaOH solution is measured potentiometrically.

In the knowledge of the total peptide content of the solutions, the quantities of acid added, the quantities of alkali consumed, and the experimental millivolt values can be used to calculate the Bjerrum-type protonation curves of the peptides showing the average number of protons bound by the peptide, $\overline{H}$, in solutions of different pH.

The equilibrium reactions describing the system, the composition of the protonated species formed, and the protonation constants could be obtained from the experimental data by computer evaluation. The essence of this is that the experimental millivolt vs. total hydrogen ion concentration curves are simulated on the basis of models assuming the stepwise formation of proton complexes of various compositions. In the construction of the models, every reasonable combination of the proton complex could be taken into consideration, and calculations could be made to the accuracy with which the different models describe the experimental curves. In this way those few combinations were obtained that describe the system correctly from a mathematical aspect.

The following chemical stipulations can be used for the selection of a single model for every system from among the models describing the experimental data with an accuracy within the experimental error over the entire concentration range:

1. With decreasing pH the value of a stepwise protonation constant must always be less than the value of the preceding constant to an extent corresponding to the statistical case at least.

2. If the difference between the constants for the successive coordination of two protons is less than that calculated from the statistical model, then the two protons must be considered as coordinating in one step.

3. From the models describing the experimental curve with the same error, the one must always be favored that is based on the assumption of the formation of fewer complexes.

In this way the protonation constants of the polyfunctional macro peptide ligands could be determined; however, the assignment of the constants to the specific functional groups remains a difficult task. This assignment could be assisted by the comparison of the corresponding protonation constants of various smaller fragments of the same macromolecule and also by their comparison with the data belonging to the functional groups of the corresponding free trifunctional amino acids.

A successful assignment of the constants to the functional groups can make possible the establishment of the positions of the hydrogen bridges influencing or even determining the conformation of the macromolecule.

2.1. The Protonation of Corticotropin Fragments in Aqueous Solutions

Burger et al. [13] determined the protonation constants of four synthetic corticotropin fragments consisting of the first N-terminal 32, 28, 14, and 4 amino acids of α_H-corticotropin ($ACTH_{1-32}$, $ACTH_{1-28}$, $ACTH_{1-14}$, and $ACTH_{1-4}$) using the procedure outlined above. From the amino acid sequence of the peptides (see Fig. 1) it can be seen that the following functional groups of ACTH can be protonated or deprotonated in the pH interval of 2-12:[*] the phenolic hydroxy groups of the tyrosines, the primary amino groups of the lysines, one of the imidazole nitrogens of the histidine, the carboxyl groups of the aspartic and glutamic acids, and the terminal amino and carboxyl groups. $ACTH_{1-32}$, $ACTH_{1-28}$, $ACTH_{1-14}$, and $ACTH_{1-4}$ possess 13, 11, 6, and 3 of these groups, respectively.

The Bjerrum-type protonation curves constructed from the experimental data are shown in Fig. 4. The protonation constants of the four peptides calculated by the computer analysis of the data

---

[*]The guanidino groups of the arginines are not considered to play a role in the deprotonation process below pH 12.

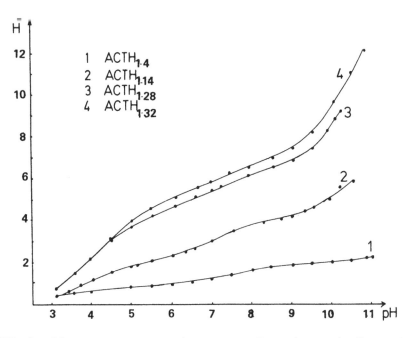

FIG. 4. Bjerrum-type protonation curves of corticotropin fragments.

are collected in Table 1. For comparison the protonation constants
of the corresponding functional groups measured in free amino acids
are also included in Table 1.

    With the help of the latter data the constants measured for
the polypeptides could be assigned with a fairly high probability
to the appropriate functional groups. The only uncertain differ-
entiation was that between the phenolic oxygens of the tyrosines
and the primary amino groups of the lysines. Because of the almost
identical basicities of these two donors, it is not possible to
exclude the overlapping of these two protonation processes and the
resulting error.

    Following the deprotonation of the phenolic hydroxy groups by
UV spectrophotometric measurements it was shown that the phenolic
hydroxy oxygen is more basic in $ACTH_{1-28}$ and $ACTH_{1-32}$ than would be
expected from the data for free tyrosine, whereas the nitrogens of
the $\varepsilon\text{-}NH_2$ groups of the lysines are more acidic than would be

TABLE 1

Logarithms of Protonation Constants of ACTH
Fragments and Individual Amino Acids

Functional group	$ACTH_{1-4}$	$ACTH_{1-14}$	$ACTH_{1-28}$	$ACTH_{1-32}$	Amino acid
Tyrosine OH	K_1 10.7	K_1 9.65	K_1K_2 19.89	$K_1K_2K_3$ 29.31	10.43
Lysine NH_2		K_2 9.45	K_3K_4 17.92	K_4 8.87 K_5 8.61	10.47
Terminal NH_2	K_2 7.17	K_3 7.10	K_5 6.97	K_6K_7 12.80	9.12
Histidine imidazole N		K_4 6.35	K_6 6.53		6.12
Glutamic acid + Aspartic acid COOH		K_5 4.24	K_7 4.95 K_8 4.59	K_8 4.64 K_9K_{10} 8.74	4.18 3.71
Terminal COOH	K_3 3.20	K_6 3.71	K_9 3.73	$K_{11}K_{12}$ 5.38	2.39

expected from the protonation constant of free lysine. On the basis
of these considerations, the assignment of the constants in Table 1
to the functional groups can be considered to be basically correct.

Unfortunately, quantitative evaluation of the UV spectral
change due to deprotonation of the two tyrosine phenolic hydroxy
groups is made uncertain by other aromatic rings (phenylalanine,
tryptophan, and histidine) of the macromolecular polypeptides.
Therefore the UV data could not be used for the calculation of
exact equilibrium data.

The data in Table 1 reflect the effect of peptide formation
on the protonation constants of the functional groups of amino acids.

The development of a peptide bond between two amino acids
separates the carboxy group from the protonated amino group, which
is more strongly electrophilic than the peptide bond; this leads to
an increase in the electron density on the oxygen of the carboxy
group and hence to an increase in the protonation constant. In
addition, the peptide group, a stronger electron acceptor than the
carboxy group, decreases the electron density on the nitrogen of the
primary amino group, which is manifested in the decrease of this
protonation constant. This double effect can be well observed if
the equilibrium constants for $ACTH_{1-4}$ are compared with the corresponding data for the free amino acids.

The changes occurring on the functional groups of the side
chains of the macromolecular polypeptide cannot be interpreted
simply by the electron shifts accompanying the formation of the
peptide bond. It can be seen from the data of Table 1 that with the
increase of the size of the peptide the basicity of the individual
ϵ-NH_2 groups decreases (see K_2 for $ACTH_{1-14}$, K_3K_4 for $ACTH_{1-28}$, and
the K_4 and K_5 values for $ACTH_{1-32}$). The basicity of the terminal
amino group decreases similarly in the above sequence. In contrast,
the basicity of the imidazole nitrogen of the histidine increases
with an increase in the length of the peptide chain (see K_4 for
$ACTH_{1-14}$, K_6 for $ACTH_{1-28}$, and K_6K_7 values for $ACTH_{1-32}$). As a
result of the opposite changes in the basicities of the terminal

amino group and the histidine nitrogen, the two protonation processes merge in the largest peptide, $ACTH_{1-32}$.

The above changes in the protonation constants of the functional groups on the side chains of the polypeptides can be explained by the formation of intramolecular hydrogen bonds.

If two protonated donor atoms, e.g., A and B, are arranged in a macromolecule in such a way that after dissociation of the proton of the less basic one (A) the two donor atoms are connected by the proton of the originally more basic donor atom (B):

$$H{-\!\!-}A + H{-\!\!-}B \rightleftharpoons H^+ + {}^-A\bullet\bullet\bullet H{-\!\!-}B$$

then, due to the formation of this hydrogen bond, dissociation of the A-H group is favored and that of B-H hindered; thus the protonation constant of A becomes smaller and that of B larger than would be in the absence of the hydrogen bridge.

On this basis it was stated that deprotonation of the two ε-NH_2 groups of $ACTH_{1-28}$ and the terminal amino group and two ε-NH_2 groups of $ACTH_{1-32}$ is promoted by the formation of hydrogen bonds with protonated donor atoms of higher basicity. Only the phenolic oxygen of the tyrosines or the guanidino nitrogen of the arginines may act as the other electronegative atom in these hydrogen bonds.

In contrast, deprotonation of the histidine imidazole nitrogen of $ACTH_{1-32}$ is inhibited by the formation of a hydrogen bond with a donor atom of lower basicity (in all probability carboxylate oxygen).

By comparison of the above findings with the conformation of ACTH (Fig. 5) suggested by Löw et al. [10] on the basis of theoretical calculations and CD measurements, the positions of the intramolecular hydrogen bridges may be suggested.

Naturally, the hydrogen bonds are formed only at a pH ensuring dissociation of the proton of the less basic donor atom, while they split at a pH causing the dissociation of the proton of the more basic donor. It thus appears that in solutions of pH > 7 the terminal amino group of the molecule may form a hydrogen bond with the phenolic hydroxy group of tyrosine (23). At pH ~9 the phenolic hydroxy group

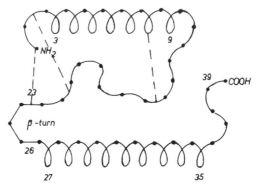

FIG. 5. The assumed secondary structure of corticotropin$_{1-32}$ based on the investigations of Löw et al. [10].

of tyrosine (2) may be linked to the primary amino group of the lysine (21), and the protonated nitrogen of the guanidino group of the arginine (8) to the primary amino group of the lysine (15). Even in aqueous solution these three hydrogen bonds may maintain the peptide molecule in a conformation that favors helix formation in a system containing less water (in solutions containing trifluoro-ethanol).

In systems of pH between 3 and 6 there is no possibility for the formation of these three hydrogen bonds. Here the protonated imidazole nitrogen of the histidine seems to be linked to the oxygen of one of the carboxylates of the molecule.

On the above basis it appears probable that while the ACTH molecule is in a conformation similar to that to be seen in Fig. 5 in alkaline aqueous solution, at about pH 6 it undergoes a conformational change.

To illustrate the agreement between the deprotonation studies and the conformation proposed by Löw et al. [10], the hydrogen bonds assumed on the basis of former investigations are indicated in Fig. 5 by dotted lines.

2.2. The Effect of the Solvent on the Protonation of Corticotropin Fragments

Circular dichroism spectra of $ACTH_{1-32}$ in aqueous solution and specially the pH dependence of these spectra indicated a relatively ordered backbone conformation of the peptide [11]. This is reflected by the protonation constants discussed in the previous section [13]. According to these two types of investigations $ACTH_{1-32}$ possesses even in water a local secondary structure probably due to the not-well-defined extended helix suggested by Tiffany and Krimm [32] for other peptides. When replacing the water in the solvent gradually by trifluoroethanol $ACTH_{1-32}$ undergoes a conformational change which leads to an α helix, probably because this solvent induces or rather reinforces interactions in the backbone and/or in the side chains.

Since the hormone must have a well-defined conformation to be recognized by its target, one would like to know which is the biologically active conformation of corticotropin. Urry et al. [33] suggested and Fermandjian et al. [34] have shown in the case of angiotensin II that the results of the experiments carried out in trifluoroethanol are similar to those obtained under physiological conditions.

Keeping in mind the idea, outlined in the previous sections, that conformation changes of the macromolecular peptides can be reflected by their protonation equilibria the effect of the change in solvent composition on the protonation processes might be of interest. Synthetic $ACTH_{1-32}$ and $ACTH_{1-4}$ served as model substances, 50 v/v % trifluoroethanol and 50 v/v % propylene-1,2-glycol as solvent in the first investigations of this type in the corticotropin system [35].

The substitution of water by water/organic solvent mixtures results in changes in the autoprotolysis constant of water (K_w). The dielectric constant of the solvent mixtures must be definitely different (mostly lower) from that of water. These differences might drastically change also the protonation equilibria. Besides all these

TABLE 2

The Solvent Dependence of the Protonation
Constants of $ACTH_{1-4}$

Protonation constant (functional group)	Solvent (dielectric constant)		
	Water (81)	50 v/v % trifluoroethanol (55)	50 v/v % propylene glycol (59)
Log K_1 (phenolic OH)	10.7	9.95	10.04
Log K_2 (terminal NH_2)	7.17	6.77	6.87
Log K_3 (COOH)	3.20	3.88	3.84

the diffusion potential difference in the solvent mixtures and in
water might cause further difficulties in the evaluation of even the
best experimental data.

Nevertheless, to get equilibrium constants in the solvent mixtures comparable with those measured in water from the potentiometric
titration curves of perchloric acid with sodium hydroxide in solutions
of composition analogous to those used for the protonation studies of
the peptides, the E_0, j_a, j_b, and K_w parameters of the equation [31]:

$$E = E_0 + 0.059 \log[H^+] + j_a[H^+] + j_b K_w [H^+]^{-1} \qquad (1)$$

were calculated, where j_a and j_b are the constants for the correction
of the effect of the diffusion potential in the acidic and basic range
of titration, respectively. For these calculations a simple computer
program is used. The good fit of the calculated points to the experimental curve is the proof of the accuracy of the parameters and indicate the millivolt range in which the experimental work could serve
data suitable for the evaluation of the equilibrium processes.

In the knowledge of the constants in Eq. (1), from the experimental data, total concentrations and millivolt values determined accordin
to the principles of classical equilibrium chemistry [30, 31], one can
calculate (using a suitable computer program) the protonation constants

of the peptides. The resulting constants are shown in Tables 2 and 3.

The protonation constants of the tetrapeptide $ACTH_{1-4}$ (Table 2) reflect the primary solvation effect due to the direct interactions of the functional groups with the solvent. In this system no secondary effect, e.g., intramolecular hydrogen bridge formation, influencing the protonation equilibria can be supposed.

It is known that the decrease of the dielectric constant of the solvent favors the association processes in the solution. The dielectric constant of both of the solvent mixtures is lower than that of water (50 v/v % trifluoroethanol $\varepsilon = 55.0$, 50 v/v % propylene $\varepsilon = 59.0$); nevertheless, only the protonation constant of the carboxylate group increased substituting the water by the solvent mixtures, the corresponding constants of the phenolic hydroxy and the primary amino groups showed a significant decrease. Thus the deprotonation of these latter groups is favored by their interaction with the organic components of the solvent mixtures.

The explanation of this behavior can be based on the strong hydrogen bridge-forming ability of trifluoroethanol and propylene glycol, respectively. In the case of the former molecule the strong electrophilic effect of the fluorines increases the availability of the alcoholic proton for the bridge formation. Propylene glycol might form a chelate-type solvate with proton acceptor groups, the chelate effect increasing the stability of the solvate. Thus both solvents might solvate the phenolic hydroxy and the primary amino group of the peptide stronger than water hindering their protonation, which is manifested in the decrease of the protonation constants.

The solvent dependence of the protonation constants of the macromolecular $ACTH_{1-32}$ was found to be different (see Table 3). Only the constants measured in the trifluoroethanol-containing solution followed the trend shown by the protonation constants of $ACTH_{1-4}$, i.e., increasing basicity for the carboxylates and decreasing basicity for the other groups. In the solution made with propylene glycol all protonation constants were found higher than in

TABLE 3

The Solvent Dependence of the Protonation
Constants (log K values) of $ACTH_{1-32}$

			Solvent	
		Water	50 v/v % trifluoro-ethanol	50 v/v % propylene glycol
Phenolic OH				10.25
				10.10
Primary NH_2	$K_1K_2K_3$	29.31		9.90 (29.6)
			9.50	9.52
	K_4	8.87		9.20
	K_5	8.61	6.74	8.65
Terminal NH_2			6.20	6.95
	K_6K_7	12.80————	——(11.91)——	————(12.91)
Hystidine N			5.71	5.96
COO^-	K_8	4.64	5.04	5.62
			4.80	4.87
	K_9K_{10}	8.74	(9.24)	(9.41)
			4.44	4.60
			4.01	4.12
	$K_{11}K_{12}$	5.38	(7.94)	(8.10)
			3.93	3.98

aqueous solutions. Since in aqueous solution some functional groups
were found equivalent and so only the products of the constants could
be determined, for comparison the analogous products are formed from
the successive constants measured in the solvent mixtures and presented in brackets in Table 3.

The different behavior of $ACTH_{1-32}$ in the two different solvent
mixtures could be explained on the basis of the different space requirement of the two organic components taking into consideration the

somewhat rigid ordered conformation of the macromolecular peptide.
The strong chelate-type solvation of the functional groups by propylene glycol is only possible when both hydroxides of the solvent can
interact with the same groups. The rigid structure of the peptide,
which is favored by the lower dielectric constants and lower water
content of the solvent mixtures, hinders this interaction. Because
of the lower space requirement of trifluoroethanol its solvation
effect prevails in the system.

Thus the effect of the solvent on the protonation equilibria
of corticotropin reflects in an indirect way the solvent dependence
of the conformation of the peptide.

2.3. Protonation Equilibria of BPTI
(Kunitz Base)

X-ray diffraction studies [20, 21] of the basic trypsin inhibitor
BPTI (Kunitz base) polypeptide show the intramolecular hydrogen
bridges in the crystalline peptide. A great part of these H bonds
are formed between the main chain peptide nitrogens and corresponding peptide oxygens; rather few H bridges connect main chain peptide
nitrogens or oxygens with side chain donor groups. Only three of
these latter type of H bonds contain donor groups with protonation
constants low enough to be studied by pH metric equilibrium measurements in aqueous solutions. These intramolecular H bridges are:

    Tyr(10)-OH···Val(34)-CO,
    Tyr(35)-OH···Cys(38)-NH   and
    Arg(1)-NH···Cys(55)-CO

bridges with bond distances 0.3, 0.3, and 0.28 nm, respectively.

The X-ray investigations did not reflect the effect of intermolecular hydrogen bridges in this system. Molecular weight determinations have shown, however, that BPTI dimerizes in solutions of
pH below 8 [36]. One can assume that this dimerization is due to
intermolecular hydrogen bridge formation between two peptide

molecules in the solution. Both intra- and intermolecular H-bridge
formation influences the protonation constants of the donor atoms
participating in these bonds. It is probable therefore that the
pH metric equilibrium study of the protonation equilibria of BPTI
reflects the effect of both types of H bonds. On the basis of this
type of investigations not only the exact pH range of the dimerization process can be determined but the assignment of the functional
groups taking part in the dimer formation could be attempted. For
this purpose the protonation study of BPTI in aqueous solutions
using the precision pH metric method outlined in Sec. 2.1 was
performed [27].

The computer evaluation of the experimental data has shown
that the system contains differently protonated monomeric and dimeric
species. Starting from the completely protonated monomer the pH
increase resulting in the dissociation of the first carboxylic acid
proton is accompanied by the formation of the dimer. A further pH
increase results first in the successive deprotonation of the dimer,
later at about pH 7 when the last protonated carboxyl group releases
its proton the dimer dissociates and the successive deprotonation of
the monomer peptide starts. The composition of the protonated species
(bound proton-peptide ratios), the protonation equilibrium constants
and the dimer formation constant are shown in Table 4. In the knowledge of the total peptide concentration and pH of the solutions the
concentration of each species can be calculated with the help of
these constants.

As a result of the protonation study of BPTI it became clear
that the deprotonation of the terminal carboxyl group and the protonation of all other carboxyl and that of the amino groups is essential for the formation of the BPTI dimer in aqueous solution. However, it is impossible to state on the basis of only equilibrium
measurements the exact place of the hydrogen bridges in the dimer.

The carboxylate oxygen can be bound by an H bond to any protonated group that has a basicity high enough to prevent its deprotonation in the pH range investigated. Such functional groups could

TABLE 4

The Protonation Constants of BPTI

Proton-peptide ratio	Successive protonation constants		Assigned functional groups
2:1	$\log K_{1.1}K_{2.1}$	22.28	Phenolic OH
4:1	$\log K_{3.1}K_{4.1}$	21.16	Phenolic OH
6:1	$\log K_{5.1}K_{6.1}$	19.88	ε-NH$_2$
7:1	$\log K_{7.1}$	9.47	ε-NH$_2$
8:1	$\log K_{8.1}$	8.97	ε-NH$_2$
9:1	$\log K_{9.1}$	7.90	terminal NH$_2$
19:2	$\log K^*$	8.27	Dimer formation*
20:2	$\log K_{20.2}$	4.36	-COOH
22:2	$\log K_{21.2}K_{22.2}$	8.71	-COOH
24:2	$\log K_{23.2}K_{24.2}$	7.41	-COOH
27:2	$\log K_{25.2}K_{26.2}K_{27.2}$	9.94	-COOH

Note: Standard deviation: 1.88 mV. $K^* = \dfrac{[H_{19}P_2]}{[H_9P][H_{10}P]}$

be the protonated primary amino groups, arginin guanidino groups and any peptide NH group. Similarly the protonated carboxyl and amino groups can form H bridges not only with the carboxylate but with any donor atom having such a low basicity that it remains unprotonated in the pH range of the investigations. Such groups could be, e.g., the peptide oxygens.

These investigations have proved unambiguously, however, that the dimerization of BPTI is due to the formation of intermolecular hydrogen bridges, and the pH dependence of this process is determined by the protonation-deprotonation equilibria of the peptide.

The results of these investigations are also the basis for the study of the metal complex formation equilibria of BPTI.

3. COMPLEX FORMATION EQUILIBRIA

Equilibrium studies of the metal complex formation reactions of polypeptides or even proteins can be based on the potentiometric determination of the free (uncomplexed) metal ion activity in solutions
containing the peptide and the metal ion in different concentrations.
For the measurement of the uncomplexed metal ion concentration in
the presence of the metal coordinated by the peptide or protein,
metal electrodes (e.g., of silver or mercury), amalgam electrodes
(e.g., zinc, cadmium, bismuth), and ion-selective membrane or ion
exchange electrodes (e.g., calcium) can be used. The coordination
of metal ions to donor atoms protonated at the pH of the solutions
to be investigated can be followed by the pH metric equilibrium
study of the deprotonation equilibria caused by this process.

Because of the large number of donor groups in such macromolecular ligands only the computer evaluation of the experimental
data can give results characterizing reliably the system. Using a
high-velocity computer program every reasonable combination of the
complex compositions (metal-ligand ratios) could be taken into consideration and the errors with which the various models describe
the experimental data, e.g., the millivolt concentration curves,
could be calculated. From among the few models describing the
measured values with an accuracy within the experimental error over
the entire concentration interval, on the basis of independent chemical information, e.g., spectroscopic studies, one might find the
right one.

Such a study of metal-polypeptide or metal-protein interactions
could lead to the determination of the number of binding sites and to
the characterization of the bonding strength of these sites by equilibrium constants. The direct assignment of the donor groups acting
as binding sites in the equilibrium processes is, however, on the
basis of only equilibrium measurements extremely difficult, in many
cases even impossible. Attempts have been made to overcome this
difficulty by comparing the stability constants for the complexes of

the macromolecules with analogous constants of smaller model ligands
with the same metal ion.

Because of the large number of possible coordination sites,
however, even such investigations may lead to uncertain or even
erroneous results. Metal ion binding by a donor group of a protein
or a macromolecular polypeptide may differ namely in several respects
from metal binding of the same group in smaller peptides.

In the macromolecule functional groups separated by many amino
acid residues may be brought close together by the tertiary structure
of the polypeptide or protein to coordinate to the same metal ion.
In this way unusual great chelate-type rings may be formed. Such
interactions are difficult to reproduce in small model compounds.

Active sites suitable for the binding of metal ions frequently
lie in clefts or pockets in the structure of the macromolecule. The
coordination of metal ions by such donor groups is extremely influenced by sterical factors, and also by the change of the solvent
structure due to the near neighbors in the pocket, e.g., the dielectric constant of water enclosed in hydrophobic pockets must be very
different from that of aqueous electrolyte solutions in which most
metal complexes have been studied. Hence, it is understandable that
the study of small model systems can give only in certain excep-
tionally favorable circumstances information on the binding sites
in the macromolecules.

In systems containing donor groups of different basicities
one might determine the binding sites in the macromolecular ligand
on the basis of the pH dependence of the equilibrium processes.
The combination of different experimental methods, e.g., comple-
menting the information deduced from the equilibrium data by spectroscopic investigations, may also help in the assignment of the
equilibrium constants to the donor groups.

In conclusion, one has to consider in general the many sided
direct investigation of the macromolecular system more promising
than the comparison of several more or less analogous model systems,
but in favorable cases such comparisons can also contribute to the

understanding of coordination processes in macromolecular systems.
The works presented in the following sections reflect the results of
both types of investigations.

3.1. Formation of Corticotropin$_{1-32}$-Zn^{2+} Complexes

Polarographic investigations [37] demonstrated that in aqueous
solution Zn^{2+} is coordinated to the functional groups of ACTH forming
successively several complexes of different stabilities. However,
because of the irreversibility of the electrode process, the compositions of these could not be established by this method with cer-
tainty. An exact potentiometric equilibrium study was therefore
made to determine the compositions and stability constants of the
complexes and to make an attempt for the assignment of the functional groups to the various successively formed complexes [14].

A zinc amalgam electrode separated by a dialysis membrane from
the solution to prevent poisoning by the macromolecular peptide was
used for the determination of the zinc ion activity change due to
complex formation [14]. From the EMF (electromotiv force) measurements and the knowledge of the total zinc ion and peptide concentrations the compositions of the species formed and the formation constants could be calculated by a suitable computer method characterized in the previous section.

This evaluation showed that the experimental curves are
described perfectly by a model assuming formation of the following
five complexes:

$$Zn_x(ACTH_{1-32})$$

where x = 2, 6, 8, 10, and 13.

The conditional stability constants at pH 5.9 and the donor
groups presumed to be involved are listed in Table 5.

Independent polarographic and deprotonation equilibrium
studies and the pH metrically determined protonation constants of

TABLE 5

Conditional Complex Products and Stepwise Stability
Constants of the Zinc-$ACTH_{1-32}$ Complexes at pH 5.9
and Assumed Assignments of the Functional Groups

Complex	$\log \beta_n$		$\log K_n$	Functional group
$Zn_2(ACTH_{1-32})$	8.93	K_1K_2	8.93	Histidine N
$Zn_6(ACTH_{1-32})$	25.48	$K_3K_4K_5K_6$	16.55	COO^-
$Zn_8(ACTH_{1-32})$	32.84	K_7K_8	7.36	Thioether S, terminal NH_2
$Zn_{10}(ACTH_{1-32})$	37.69	K_9K_{10}	4.85	ε-NH_2
$Zn_{13}(ACTH_{1-32})$	44.34	$K_{11}K_{12}K_{13}$	6.65	Arginine N

Note: Standard deviation: 1.85 mV.

corticotropin fragments assisted in the attempt to assign the
stability constants to the functional groups.

The deprotonation experiments showed that in the pH range
4.0 to 6.5 with even a 100-fold zinc excess the coordination of
Zn^{2+} results in the expulsion of not more than three protons from
one $ACTH_{1-32}$ molecule.

On the other hand, the Bjerrum-type formation curve of the
zinc complex of $ACTH_{1-32}$ constructed from the data of the zinc
amalgam electrode measurements at pH = 5.9 clearly demonstrates
that one $ACTH_{1-32}$ molecule binds a maximum of 13 zinc ions.

These two independent examinations prove that the majority of
the zinc ions are coordinated to donor atoms which are not protonated
at the pH employed (pH 5.9). Such donor atoms are the five carboxylate oxygens, the thioether sulfur atom of the methionine, one of the
nitrogens of the imidazole ring of histidine, and one nitrogen of
the guanidino groups of each of the three arginines, i.e., a total
of 10 functional groups.

The deprotonation investigations indicate that at the given
pH only three zinc ions are connected to donor atoms that were
originally protonated. Such donor atoms are the nitrogens of the

five primary amino groups, the oxygens of the two phenolic hydroxy
groups, the other nitrogen of the imidazole ring of the histidine,
and the other nitrogens of the guanidino groups of the three arginines, i.e., a total of 11 donor atoms.

The polarographic study of $ACTH_{1-32}$ and its zinc complexes has
shown that ACTH catalyzes the hydrogen discharge on the mercury
electrode. However, this catalytic effect is lower within a given
pH range than the one measured in the presence of excess zinc.
This indicates that zinc ions are coordinated with ACTH donor atoms
whose protons (or the proton of the neighboring atom which can be
in conjugate position) become loosened, thus bringing about a catalytic hydrogen evolution.

The pH dependence of this latter catalytic hydrogen wave
clearly demonstrated that in the complex Zn^{2+} is coordinated, among
others, to the nonprotonated nitrogen of the histidine imidazole
ring, and thus catalyzes reduction of the hydrogen of the protonated
ring nitrogen.

This catalytic effect appears in systems containing an $ACTH_{1-32}$
excess, indicating that the zinc ion is coordinated to the corresponding histidine nitrogen in Zn_2ACTH, the first successive complex
formed at low zinc concentration.

From the zinc concentration dependence of the deprotonation
process it could be seen that formation of the first two complexes
Zn_2ACTH and Zn_6ACTH is not accompanied by deprotonation. In the
coordination of the first six zinc ions, therefore, the five carboxylate groups, which are in deprotonated form at pH 5.9, may take
part besides the nonprotonated histidine nitrogen.

Calculations based on the zinc concentration dependence of the
deprotonation processes has indicated also that the release of the
three protons caused by the coordination of zinc ions can be attributed to the formation of the Zn_8ACTH and $Zn_{10}ACTH$ species. From the
protonation constant sequence of ACTH it may be assumed that the
deprotonation in these processes occurs on the terminal amino group
and on two lysine amino groups, the fourth Zn^{2+} bound during these

steps of complex formation must be coordinated again to a nonprotonated donor atom, probably the thioether sulfur.

Since formation of the final complex is again not accompanied
by proton substitution, by consideration of what nonprotonated donor
atoms remain in the polypeptide molecule it may be assumed that the
nonprotonated nitrogens of the guanidino groups of the three arginines or probably peptide oxygens participate in the formation of the
complex $Zn_{13}(ACTH_{1-32})$.

In the above interpretation it was assumed that only one of
the $ACTH_{1-32}$ donor atoms is coordinated to each bound zinc ion, and
that the other sites in the coordination sphere of Zn^{2+} are occupied
by water molecules. This assumption is supported by the good water
solubility of the macromolecular zinc complexes.

These considerations do not exclude the possibility that in
complexes with low zinc contents, and primarily in $Zn_2(ACTH_{1-32})$,
donor groups originally deprotonated at pH 5.9 are linked to other
coordination sites of the Zn^{2+}. Increase of the zinc ion concentration in the system may cause the rearrangement of the complexes
initially formed.

Even such a short survey of the results indicate, that all
these suggestions for the assignment of the coordination sites have
an uncertainty. Probably the effect of the coordination of zinc
ions on the proton resonance spectra of ACTH could help in the
unambiguous determination of the binding sites.

3.2. The Silver Complex Formation of
       Corticotropin Fragments

The potential complex-forming functional groups of ACTH contain
oxygen, nitrogen, or sulfur donor atoms. Some of them are strongly
basic, i.e., the guanidino and primary amino groups, which are
protonated in a wide pH range and therefore their metal ion coordination results in the substitution of protons. On the other hand,
the complex formation of the less basic donor atoms, i.e., of the

carboxylate groups and the methionine thioether sulfur, is pH independent in a wide concentration range.

It is known that the functional groups have a different affinity towards each metal ion. Thus, according to Pearson [38], the hard carboxylate oxygen preferably links to the hard alkali- and alkali-earth metal ions, whereas the soft sulfur donor atoms do so to the soft silver, mercury, and platinum ions. The nitrogen donor atoms are on the border line.

The aim of the equilibrium study of the silver complex formation of the corticotropin fragments $ACTH_{1-32}$, $ACTH_{1-28}$, and $ACTH_{1-4}$ was to unravel the coordination chemical behavior of the methionine sulfur in the macromolecular peptide [15].

Since some donor atoms of ACTH are protonated and others nonprotonated in the pH range of the studies, the silver complex formation is composed of pH-independent and pH-dependent steps. To separate these processes in the course of the equilibrium measurements not only the silver ion activity but also the pH of the solution was followed, and the pH change due to Ag^+ coordination was compensated by using a sodium hydroxide standard solution. A three-electrode potentiometric method using a silver electrode for the measurement of the silver ion activity, a glass electrode for the pH measurement, and a reference electrode (silver-silver chloride) placed in a Wilhelm bridge [39] served this purpose.

The equilibrium constants were calculated from the experimental data on the basis of various suggested models, e.g., Ag^+-to-peptide ratios. The analytical concentrations (total peptide and silver ion concentration), silver ion activity, pH, and the sodium hydroxide amount needed to keep the pH constant served as primary experimental data.

The equilibrium measurements have shown that increasing the silver ion concentration in the peptide-containing solutions the coordination of the first silver ion is well separated from the successive binding of the following silver ions. Therefore this former process was thoroughly investigated.

Comparing the corresponding equilibrium data of the different corticotropin fragments measured in the concentration range of the coordination of the first silver ion a striking similarity could be observed, indicating the analogy of these processes in the ACTH fragments of different size. Thus this silver ion must be bound by identical functional groups in the different ACTH fragments. Since ACTH$_{1-4}$, the N-terminal four amino acid containing corticotropin fragment was the smallest among the peptides investigated, the donor groups situated in this part of the macromolecules must be considered to take part in the coordination process. These are the terminal amino and the tyrosine-2 phenolic hydroxy group and the methionine (4) thioether sulfur donor atom.

The pH dependence of the coordination of the first silver ion, measured by the silver electrode in solutions of different but constant pH and the amount of protons released in the course of complex formation, indicated the overlapping of two equilibria: (a) the pH-independent coordination of silver ion to the methionine (4) sulfur; in this process the peptide seems to act as a monofunctional ligand; (b) the pH-dependent coordination of the terminal amino group by the silver bound in the previous step to the methionine residue of the peptide (the closing of a chelate ring of rather unusual great size favored by the linear sp hybridization of the coordinated silver).

In solutions of pH $\leq$ 3 only process (a) described by equation

$$Ag^+ + HP \rightleftharpoons AgHP^+ \tag{a}$$

takes place. A further increase of the pH results in the appearance of the pH-dependent process (b):

$$AgHP^+ \rightleftharpoons AgP + H^+ \tag{b}$$

The bonding strength of silver in the complex species AgHP$^+$ and AgP, respectively, is characterized by the stability constants:

$$K_S = \frac{[AgHP^+]}{[Ag^+][HP]} \quad \text{and} \quad K_N = \frac{[AgP]}{[Ag^+][P^-]}$$

shown in Table 6.

TABLE 6

Stability Constants of the Silver Complexes
of ACTH Fragments

| Peptide | pH | $\log K_S$ | | | $\log K_N$ |
		1	2	3	5
$ACTH_{1-4}$		3.50 ± 0.04	3.64 ± 0.04	3.56 ± 0.05	5.10 ± 0.1
$ACTH_{1-28}$		3.65 ± 0.1		3.63 ± 0.05	5.50 ± 0.1
$ACTH_{1-32}$		3.72 ± 0.1		3.58 ± 0.04	5.22 ± 0.1

The concentration $[P^-]$ was calculated from the total peptide concentration using the protonation constants of the terminal amino groups of the corresponding peptides. The concentration and pH independence of the K_N values is the clear-cut proof of the correctness of the assignment of the terminal amino groups in process (b).

To support the suggestion of the assignment of the thioether sulfur atom in both processes (a) and (b), in one of the peptides, i.e., in $ACTH_{1-28}$, this sulfur was oxidized to a sulfoxy group preventing in this way its coordination to Ag^+. The silver complex formation study of this oxidized peptide reflected a dramatic decrease in the stability of the Ag^+ complex, proving in this way the correctness of the assignment.

The great difference in the values of the stability constants K_N and K_S ($K_N > K_S$) supports the assumption that the peptide acts in process (a) as a monofunctional ligand, and the chelate effect caused the increased stability of the species formed in process (b).

According to the data in Table 6, both constants are independent of the size of the peptide indicating that those parts of the molecule that do not interact directly with the coordinated silver have no effect on the stability of the complexes.

3.3. The Silver Complex Formation
of BPTI (Kunitz Base)

Sulfur atoms have great importance in the determination of the structure and chemical behavior of the basic trypsin inhibitor (BPTI) polypeptide. The exceptional conformational stability of this macromolecule is secured by three disulfide bridges. Two of them, cys(5)-cys(55) and cys(30)-cys(51), are completely buried within the molecule, and the third one, cys(14)-cys(38), seems to be accessible from the outside. The sulfur atoms of this last group and the thioether sulfur of methionine(52) can be considered to be the Pearson-type [38] soft centers of the molecule.

In spite of this some organomercury compounds are bound by different nitrogen donor atoms, i.e., amino and amido groups, of the molecule [20]. Since the soft acceptor character of mercury favoring its bonding by soft donor groups could be changed by the organic moiety in the aromatic mercury compounds, the accessibility of the soft sulfur donor atoms in coordination reactions could be better studied following the silver complex formation of BPTI [28] by potentiometric equilibrium measurements.

The experimental technique discussed in Sec. 3.2 was used for the study of this system, too. Thus besides the changes in the silver ion activity, also the amount of protons released due to the coordination process was determined.

The results of these investigations have shown that the silver ion coordination taking place in the first step is not accompanied by the deprotonation of the peptide. In this (between pH 2.7 and 6.1) pH-independent step formally a maximum of half a mole of silver ion gets coordinated to one mole of peptide molecule. This process thus indicates the formation of a silver bridge linking two peptide molecules, or the dimerization of the peptide before the complex formation.

According to calculations made on the basis of the concentration dependence of the equilibrium data representing this complex
formation step, the latter suggestion is the correct one. Thus
BPTI dimerizes in solution (see also Sec. 2.3) and the dimer is
coordinated by the silver ion. In this process only donor groups
participate which are not protonated in the investigated pH range.
These are the thioether sulfur atom of methionine (52) and the cys
(14-38) disulfide, which seems to be accessible in contrast to the
other strongly overshadowed disulfide bridges of the molecule.

The X-ray structure investigation demonstrating the intermolecular interaction between peptide molecules in the crystalline
state shows that BPTI form endless rods of molecules linked head to
tail in such a way that the cys (14-38) disulfide bridge at the top
of a BPTI gets near to the methionine (52) sulfur atom of the next
peptide. Therefore a plausible suggestion is that the silver bridge
connects these two groups in the BPTI dimer-containing complex.
Indeed, the donor properties of thioether sulfur have been shown by
recent investigations of Sigel et al. [40], and the affinity of
disulfide sulfur towards Ag^+ was shown in the oxytocin silver
system by Burger and Zay [41].

A further increase in the silver ion concentration results
formally in the coordination of a second half of silver ion per one
peptide. This process is, however, accompanied by deprotonation;
half a mole of proton is released in the maximum per one mole of
peptide. Hence the BPTI dimer containing one silver, which is
bound in the first coordination step, binds an additional silver ion
in a pH-dependent process.

Computations made from the pH and concentration dependence of
the equilibrium data representing this latter process have shown
that the coordination of the second Ag^+ resulted in a system containing three species of a 2:2 silver-peptide composition. The
three complex species differ only in their degree of protonation.
The formation of the first one did not change the protonation of
the peptide. This complex releases first one and successively
afterward a second proton.

On the basis of these measurements the following four
equations reflect the Ag^+-BPTI interaction (the logarithms of
the corresponding equilibrium constants are given in brackets):

$$Ag^+ + (HP)_2 \rightleftharpoons Ag(HP)_2^+ \qquad (5.45 \pm 0.05) \qquad (1)$$

$$Ag^+ + Ag(HP)_2^+ \rightleftharpoons Ag_2(HP)_2^{2+} \qquad (2.30 \pm 0.08) \qquad (2)$$

$$Ag_2(HP)_2^{2+} \rightleftharpoons Ag_2HP_2^+ + H^+ \qquad (-3.13 \pm 0.06) \qquad (3)$$

$$Ag_2HP_2 \rightleftharpoons Ag_2P_2 + H^+ \qquad (-6.60 \pm 0.08) \qquad (4)$$

where HP denotes the BPTI molecule neglecting its charge.

In solutions of silver ion excess and in the pH range 5 to 6
$Ag_2HP_2^+$ is the dominant species. In this complex the second silver
ion can also be suggested to form a bridge between the two peptides
in the complex. But whereas the first coordinated silver ion connects two originally nonprotonated donor groups, i.e., the sulfur
donor atoms in $Ag(HP)_2^+$, this second bridge in $Ag_2HP_2^+$ is connecting
one originally nonprotonated and one in process (3) deprotonated
donor group.

The X-ray investigation reflecting the intermolecular relations shows that the glutamic acid (49) and the arginine (39) of two
neighboring peptide molecules are closely situated. Because the
carboxyl of glutamic acid is deprotonated in the investigated pH
range, the intermolecular silver bridge can be assumed to connect
these two amino acid residues. The coordination of silver in this
case results in the strikingly great increase of the acidity of the
guanidino group of arginine.

The dissociation of the next proton indicates the entry of a
new, originally protonated, functional group into the coordination
sphere.

The processes discussed above might be assumed, however, to
take place with the participation of other groups as well, e.g.,
the second silver bridge might connect the amino group of lysine (15)
of one peptide with a carboxylate group in the terminal part of the
second BPTI assuming head to tail dimerization, or the terminal
amino group of one BPTI with the carboxylate of the other assuming
the dimerization through the terminal part of both peptides.

On the basis of only equilibrium measurements it is impossible
to decide which suggestion is the right one. Probably the effect of
silver ion coordination on the nmr spectrum of BPTI could help in
the exact determination of the bonding sites.

4. CONCLUSIONS

The protonation and complex formation studies discussed in
this review reflect the possibilities and restrictions of equilibrium
methods in the characterization of macromolecular polyfunctional
ligand systems.

It is to be seen that the classical methods of equilibrium
chemistry could lead to the unambiguous determination of the number
of binding sites in the protonation and complex formation processes.
The computer evaluation of the experimental data results even in
the determination of the composition (metal-ligand and proton-ligand
ratios) of the species in the solution. The equilibrium constants
derived from the measurement characterize the bonding strength of
the donor groups in the different species.

Results of equilibrium investigations may even assist in the
assignment of the functional groups taking part in the protonation
and complex formation processes. However, on the basis of equilibrium studies alone this assignment is in most macromolecular systems
impossible, or at least extremely difficult.

The protonation equilibrium study of the macromolecular peptides ACTH and BPTI has demonstrated that this type of investigation
may reflect the effect of inter- and intramolecular hydrogen bridge
formation. Since the existence of H bridges depends on the secondary
and even tertiary structure of the polypeptides, the results of such
equilibrium studies can also give (indirectly) structural information
on the system.

ACKNOWLEDGMENT

The ACTH fragments and the BPTI peptide investigated are the products of G. Richter Chemical Works, Budapest, Hungary, to whom the author expresses his thanks.

I wish to thank also Dr. F. Gaizer and Dr. B. Noszál for helpful discussions, and Dr. B. Noszál, Dr. M. Pékli, and Dr. I. Zay for their participation in the experimental work.

REFERENCES

1. R. Österberg, in Metal Ions in Biological Systems (H. Sigel, ed.), Marcel Dekker, New York, Vol. 3, 1974, p. 45 and the references therein.

2. J. Ramachandran, in Hormone Proteins and Peptides (C. H. Li, ed.), Academic Press, New York, Vol. 2, 1973.

3. T. H. Lee, A. B. Lerner, and V. Buetner-Janusch, *J. Biol. Chem.*, *236*, 2970 (1961).

4. B. Riniker, P. Sieber, and W. Rittel, *Nature New Biol.*, *235*, 114 (1972).

5. S. Bajusz, K. Medzihradszky, Z. Paulay, and Zs. Láng, *Acta Chim. Acad. Sci. Hung.*, *52*, 335 (1967)

6. L. C. Craig, J. D. Fischer, and J. P. King, *Biochemistry*, *4*, 311 (1965).

7. H. Edelhoch and R. E. Lippoldt, *J. Biol. Chem.*, *244*, 3876 (1969).

8. C. H. Li, *Recent Progress Horm. Res.*, *18*, 1 (1962).

9. K. Hofmann and H. Yajima, *Recent Progress Horm. Res.*, *18*, 1 (1962).

10. M. Löw, L. Kisfaludy, and S. Fermandjian, *Acta Biochim. Biophys. Acad. Sci. Hung.*, *10*, 229 (1975).

11. D. Greff, F. Toma, S. Fermandjian, M. Löw, and L. Kisfaludy, *Biochim. Biophys. Acta*, *439*, 219 (1976).

12. P. G. Squire and T. Bewley, *Biochem. Biophys. Acta*, *109*, 234 (1965).

13. K. Burger, F. Gaizer, B. Noszál, M. Pékli, and G. Takácsi Nagy, *Bioinorg. Chem.*, *7*, 335 (1977).

14. K. Burger, F. Gaizer, I. Zay, M. Pékli, and B. Noszál, *J. Inorg. Nucl. Chem., 40,* 725 (1978).

15. B. Noszál and K. Burger, *Inorg. Chim. Acta,* in press.

16. J. Chauvet, G. Nouvel, and R. Archer, *Biochim. Biophys. Acta, 92,* 200 (1964).

17. B. Kassell and M. Laskovski, Jr., *Biochem. Biophys. Res. Commun., 20,* 463 (1965).

18. F. A. Anderer and S. Hörnle, *J. Biol. Chem., 241,* 1568 (1966).

19. R. Huber, D. Kukla, O. Epp, and H. Formanek, *Naturwissenschaften, 57,* 389 (1970).

20. R. Huber, D. Kukla, A. Rühlmann, and W. Steigemann, in Proc. Int. Res. Conf. Proteinase Inhibitors, Munich, Nov. 1970 (H. Fritz and H. Tschesche, eds.), De Gruyter, Berlin-New York, 1971, p. 56.

21. J. Deisenhoffer and W. Steigemann, in Bayer Symposium V. Proteinase Inhibitors (H. Fritz, H. Tschesche, L. J. Green, and E. Truscheit, eds.), Springer Verlag Berlin, Heidelberg, New York, 1974, p. 484.

22. H. Fritz, I. Trautschold, and E. Werle, *Z. Physiol. Chem., 342,* 253 (1965).

23. H. Fritz, H. Schult, R. Meister, and E. Werle, *Z. Physiol. Chem., 350,* 1531 (1969).

24. J. Pütter, *Z. Physiol. Chem., 348,* 1197 (1967).

25. A. Rühlmann, D. Kukla, P. Schwager, K. Bartels, and R. Huber, *J. Mol. Biol., 77,* 417 (1973).

26. R. Huber, D. Kukla, W. Steigemann, J. Deisenhofer, and A. Jones, in Bayer Symposium V. Proteinase Inhibitors (H. Fritz, H. Tschesche, L. J. Green, and E. Truscheit, eds.), Springer Verlag, Berlin, Heidelberg, New York, 1974, p. 497.

27. K. Burger, I. Zay, F. Gaizer, and B. Noszál, *Inorg. Chim. Acta,* in press.

28. K. Burger, I. Zay, and B. Noszál, *J. Inorg. Nucl. Chem.,* to be published.

29. K. Burger and A. Vértes, unpublished results.

30. M. T. Beck, Chemistry of Complex Equilibria, Van Nostrand, London, 1969.

31. F. J. C. Rossotti and H. Rossotti, The Determination of Stability Constants, McGraw-Hill, New York, Toronto, London, 1961.

32. M. L. Tiffany and S. Krimm, *Biopolymers, 11,* 2309 (1972).

33. D. W. Urry, L. Masotti, and J. R. Krivacic, *Biochim. Biophys. Acta, 241,* 600 (1971).

34. S. Fermandjian, D. Greff, and P. Fromageot, in Chemistry and
     Biology of Peptides (J. Meienhofer, ed.), *Ann. Arbor Sci. Publ.*,
     1972, 454-562.

35. K. Burger, B. Noszál, and F. Gaizer, unpublished results.

36. F. A. Anderer and S. Hörnle, *Z. Naturforsch., 20b,* 457 (1965).

37. K. Burger, G. Farsang, L. Ladányi, B. Noszál, M. Pékli, and
     G. Takácsi Nagy, *Bioelectrochem. Bioenerg., 2,* 329 (1975).

38. R. G. Pearson, *J. Amer. Chem. Soc., 85,* 3533 (1963).

39. W. Forsling, S. Hietanen, and L. G. Sillén, *Acta Chem. Scand.,
     6,* 905 (1952).

40. V. M. Rheinberger and H. Sigel, *Naturwissenschaften, 62,* 182
     (1975). See also D. B. McCormick, R. Griesser, and H. Sigel,
     in Metal Ions in Biological Systems (H. Sigel, ed.), Marcel
     Dekker, New York, 1974, Vol. 1, p. 213.

41. K. Burger and I. Zay, to be published.

42. B. Noszál and K. Burger, *Acta Chim. Acad. Sci. Hung.* Centenary
     Volume, 1979.

NOTE ADDED IN PROOF

The exact interpretation of the coordination chemical equilibria of
polyfunctional ligands is made rather difficult by overlapping
processes. Very recently a new method has been elaborated [42] for
the determination of equilibrium constants characterizing the single
functional groups in such systems (called group constant). The
correlation between the complex products (β values) measured as
shown in this review and the group constants was determined. For
the protonation of a three-functional base these correlations are,
e.g.:

$$\beta_1 = K_A + K_B + K_C$$

$$\beta_2 = K_A K_B + K_A K_C + K_B K_C$$

$$\beta_3 = K_A K_B K_C$$

where K_A, K_B, and K_C are the group constants characterizing the
protonation of the three basic donor groups.

It can be seen that the number of unknown group constants and that of the complex products known from the equilibrium measurements is equal. Thus the unknown group constants can be calculated by a suitable mathematical method from the known β values.

This procedure was found to be suitable for the exact evaluation of equilibrium measurements on such macromolecular polypeptides as the corticotropin fragments.

AUTHOR INDEX

Numbers in parentheses are reference numbers and indicate that an author's work is referred to although his name may not be cited in the text. Underlined numbers give the page on which the complete reference is listed.

Printed and bound by CPI Group (UK) Ltd, Croydon, CR0 4YY

17/10/2024

01775696-0011